To my parents, Luke and Reone for their unconditional love and support.
Mom, you'll be forever missed (1936 – 2010).

—*Brad*

Contents at a Glance

Pro jQuery Mobile

Brad Broulik

Apress®

Pro jQuery Mobile

President and Publisher: Paul Manning
Lead Editor: Michelle Lowman
Development Editor: Susan Ethridge
Technical Reviewer: Jorge Ramon
Editorial Board: Steve Anglin, Mark Beckner, Ewan Buckingham, Gary Cornell, Morgan Ertel, Jonathan Gennick, Jonathan Hassell, Robert Hutchinson, Michelle Lowman, James Markham, Matthew Moodie, Jeff Olson, Jeffrey Pepper, Douglas Pundick, Ben Renow-Clarke, Dominic Shakeshaft, Gwenan Spearing, Matt Wade, Tom Welsh
Coordinating Editor: Jessica Belanger
Copy Editor: Lori Cavanaugh
Compositor: MacPS, LLC
Indexer: SPi Global
Artist: SPi Global
Cover Designer: Anna Ishchenko

Distributed to the book trade worldwide by Springer Science+Business Media, LLC., 233 Spring Street, 6th Floor, New York, NY 10013. Phone 1-800-SPRINGER, fax (201) 348-4505, e-mail orders-ny@springer-sbm.com, or visit www.springeronline.com.

For information on translations, please e-mail rights@apress.com, or visit www.apress.com.

Apress and friends of ED books may be purchased in bulk for academic, corporate, or promotional use. eBook versions and licenses are also available for most titles. For more information, reference our Special Bulk Sales–eBook Licensing web page at www.apress.com/bulk-sales.

The information in this book is distributed on an "as is" basis, without warranty. Although every precaution has been taken in the preparation of this work, neither the author(s) nor Apress shall have any liability to any person or entity with respect to any loss or damage caused or alleged to be caused directly or indirectly by the information contained in this work.

The source code for this book is available to readers at www.apress.com. You will need to answer questions pertaining to this book in order to successfully download the code.

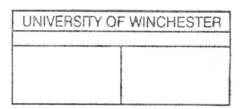

Contents

About the Author

 Brad Broulik is a senior developer specializing in enterprise mobile development at HealthPartners. Prior to mobile development he was the lead software architect at a financial services organization. He has extensive experience with most mobile technologies, particularly jQuery Mobile. He has contributed multiple pull requests to the jQuery Mobile project on GitHub and is actively developing several enterprise mobile apps with jQuery Mobile. He blogs regularly at http://bradbroulik.blogspot.com, tweets as @BradBroulik, and can be contacted at BradBroulik@gmail.com. He lives with his wife and daughter in Minnesota and enjoys exercising, outdoor activities, and spending time with family.

About the Technical Reviewer

Jorge Ramon is the founder of a small technology company that focuses on software development and developer education on mobile, web, and desktop technologies. He is also the author of the *Ext JS 3.0 Cookbook*.

Acknowledgments

I want to extend an enormous amount of appreciation to the entire jQuery Mobile core team: John Resig, Todd Parker, Scott Jehl, Kin Blas, John Bender, and Ghislain Seguin. It has been a pleasure to write about your remarkable creation.

I also want to thank the entire Apress team for their guidance and feedback: Michelle Lowman, editor; Jessica Belanger, coordinating editor; Susan Ethridge, developmental editor; Jorge Ramon, technical reviewer. This book would not have been possible without your help.

I want to especially thank my wife and daughter for their patience, support, and assistance with this project. Thank you!

Introduction

We are currently witnessing a shift in the way enterprises and individuals build and distribute mobile applications. Initially, the strategy was to build separate native apps for each major platform. However, teams quickly realized that maintaining multiple platforms was unsustainable as mobile teams lost their agility. The mobile teams that can build once and ship to all devices tomorrow will have a competitive advantage and jQuery Mobile can help get you there.

jQuery Mobile is a framework for delivering cross-platform mobile web applications with a unified interface. jQuery Mobile combines responsive layouts with progressive enhancement to render the best possible user experience from a single code base. With jQuery Mobile, we will see how to create themable, responsive, native-looking applications for iOS, Android, Windows Phone, Blackberry, and more. We will discover what sets jQuery Mobile apart from other mobile web development platforms and we will walk through practical examples of jQuery Mobile features, including design elements and event handling.

What you'll learn

- Unique features of jQuery Mobile
- jQuery Mobile core features, including page structure, navigation, form elements, lists, and grids
- How to create themable designs
- The entire jQuery Mobile API, including data attributes, methods, and events
- Integrating web services, Google Maps, and Geolocation into your jQuery Mobile apps
- How to extend jQuery Mobile with PhoneGap when you need to distribute to an app store or access device functionality
- How to apply jQuery Mobile to specific cases, including iOS and Android apps

Who This Book Is For

Mobile developers who want to master jQuery Mobile and build cross-platform mobile web applications from a single code base.

Downloading the code

The source code for this book is available to readers at http://www.apress.com in the Source Code section of the book's home page. Please feel free to visit the Apress web site and download the code there.

Chapter 1

Why jQuery Mobile?

jQuery Mobile is a new, simple to use, UI framework for building cross-platform Mobile Web applications. In a matter of minutes, you can create mobile applications (apps) that are optimized to run on nearly every phone, tablet, desktop, and e-reader device available today. That's right, with a single jQuery Mobile codebase we can create a unified experience for nearly all consumers. jQuery Mobile is an ideal framework for any Web designer or developer who needs a simple framework for creating a rich mobile Web experience. The experience also extends beyond the Web. jQuery Mobile apps can also be compiled with hybrid techniques for distribution within your favorite native app store. As we begin our journey, let's review the important features that make jQuery Mobile unique.

Universal Access

jQuery Mobile applications are universally accessible to all devices with a browser. This is a favorable reach advantage to jQuery Mobile's distribution model (see Figure 1–1). Nearly every mobile device ships with a browser. If your app is universally accessible to this broad spectrum it is a major competitive advantage. The following is a complete listing of supported devices in jQuery Mobile 1.0, which includes most phones, tablets, desktop browsers, and even e-readers.

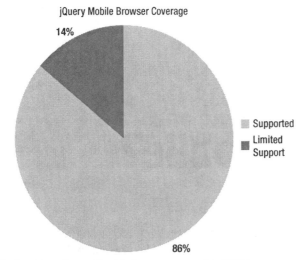

jQuery Mobile Browser Coverage

14%

Supported

Limited
Support

86%

http://gs.statcounter.com/#mobile_browser-ww-monthly-201101-
201106-bar

Figure 1–1. *jQuery 1.0 Mobile Browser Coverage*

Supported Devices:

- Phones/Tablets
  - Android 1.6+
  - BlackBerry 5+
  - iOS 3+
  - Windows Phone 7
  - WebOS 1.4+
  - Symbian (Nokia S60)
  - Firefox Mobile Opera Mobile 11+
  - Opera Mini 5+
- Desktop browsers
  - Chrome 11+
  - Firefox 3.6+
  - Internet Explorer 7+
  - Safari
- e-readers
  - Kindle
  - Nook

> **NOTE:** For an up-to-date listing of all supported platforms, refer to jQuery Mobile's supported platforms page (see http://jquerymobile.com/gbs/).

Comparatively, native application development has a very restrictive distribution model (see Figure 1–2). Native applications are only available on their native operating system. For example, an iPhone app is only accessible from an iOS device. If your goal is to reach the most consumers possible this distribution model is limited. Fortunately, jQuery Mobile apps are not restricted by this distribution barrier.

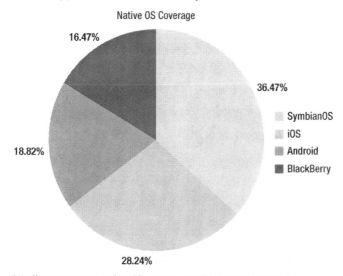

Native OS Coverage

16.47%

36.47%

SymbianOS
iOS
Android
BlackBerry

18.82%

28.24%

|http://gs.statcounter.com/#mobile_os-ww-monthly-201101-201106-bar

Figure 1–2. *Native OS Coverage*

n addition to universal access, jQuery Mobile applications can take advantage of the instant deployment capabilities we have grown accustomed to on the Web. For jQuery Mobile apps, no barriers exist in regards to certification reviews that are required within the native app distribution model. Mobile Web apps can be updated and deployed instantly to your production users. For example, I was recently working on a native enterprise application that needed an update and it took a week for the re-certification process to approve the change. In all fairness, the native app stores have options to submit urgent updates but the point is you will be dependent upon a third party to push the update to their store. The instantaneous deployment model of the Mobile Web is very advantageous in this regard.

Unified UI Across All Mobile Platforms

jQuery Mobile delivers a unified user interface by designing to HTML5 and CSS3 standards. Mobile users expect their user experience to be consistent across platforms (see Figure 1–3, Figure 1–4, Figure 1–5). Conversely, compare the native Twitter apps on both iPhone and Android. The experience is not unified. jQuery Mobile applications remedy this inconsistency, providing a user experience that is familiar and expected, regardless of the platform. Additionally, a unified user interface will provide consistent documentation, screen shots, and training regardless of the end user platform. For example, if your sales staff needs training on a new mobile app that is being deployed the user documentation will contain consistent screen shots that apply to all platforms. If half the team has iPhones and the other half has Android devices, the training experience and documentation will be the same for all users.

Figure 1–3. *iPhone*   **Figure 1–4.** *Windows Phone*   **Figure 1–5.** *Android*

jQuery Mobile also helps eliminate the need for device-specific UI customizations. A single jQuery Mobile code base will render consistently across all supported platforms without customizations. This is a very cost-effective solution compared to an organization supporting a native code base per OS. And supporting a single code base is much more cost-effective long term in regards to support and maintenance costs (see Figure 1–6).

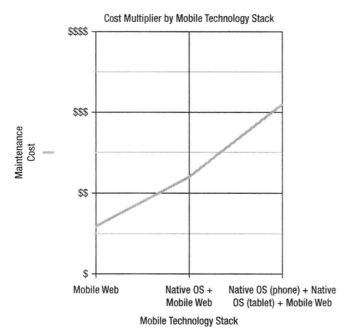

Figure 1–6. *Cost Multiplier by Mobile Technology Stack*

Simplified Markup-Driven Development

jQuery Mobile pages are styled with HTML5 markup (see Listing 1–1.). Aside from the new custom data attributes introduced in HTML5, everything should appear very familiar for Web designers and developers. If you are already familiar with HTML5 migrating to jQuery Mobile should be a relatively seamless transition. In regards to JavaScript and CSS, jQuery Mobile does all of the heavy lifting by default. However, there are instances where you may need to rely upon JavaScript to create a more dynamic or enhanced page experience. In addition to the simplicity of the markup required to design pages it also allows for rapid prototyping of user interfaces. Very quickly we can create a static workflow of functional pages, transitions, and widgets to help our customers see live prototypes with minimal effort.

Listing 1–1. *Insert listing caption here.*

```
<!DOCTYPE html>
<html>
<head>
  <meta charset="utf-8">
  <title>Title</title>
  <meta name="viewport" content="width=device-width, initial-scale=1">
  <link rel="stylesheet" href="jquery.mobile-1.0.min.css" />
  <script type="text/javascript" src="jquery-1.6.2.min.js"></script>
  <script type="text/javascript" src="jquery.mobile-1.0.min.js"></script>
</head>
<body>

<div data-role="page">
    <div data-role="header">
        <h1>Page Header</h1>
    </div>

    <div data-role="content">
        <p>Hello jQuery Mobile!</p>
    </div>

    <div data-role="footer">
        <h4>Page Footer</h4>
    </div>

</div>

</body>
</html>
```

Progressive Enhancement

jQuery Mobile will render the most elegant user experience possible for a device. For example, look at the jQuery Mobile switch control in Figure 1–7. This is the switch control on an A-Grade browser.[1]

---

[1] An A-Grade browser supports media queries and will render the best experience possible from jQuery Mobile CSS3 styling.

Figure 1–7. *A-Grade Experience*

Figure 1–8. *C-Grade Experience*

jQuery Mobile renders the control with full CSS3 styling applied. Alternatively, Figure 1–8 is the same switch control rendered on a much older C-Grade browser.[2] The C-grade browser does not render the full CSS3 styling.

> **IMPORTANT:** Although the C-Grade experience is not the most visual appealing it demonstrates the usefulness of graceful degradation. As users upgrade to newer devices, the C-Grade browser market will eventually diminish. Until this crossover takes place, C-Grade browsers will still receive a functional user experience when running a jQuery Mobile app.

Native applications do not always degrade as gracefully. In most cases, if your device does not support a native app feature you will not even be allowed to download the app. For instance, a new feature in iOS 5 is iCloud persistence. This new feature simplifies data synchronization across multiple devices. For compatibility, if we create a new iOS app that incorporates this new feature we will be required to set the "minimum allowed SDK" for our app to 5.0. Now our app will only be visible in the App Store to users running iOS 5.0 or greater. jQuery Mobile applications are more flexible in this regard.

---

[2] A C-Grade browser does not support media queries and will not receive styling enhancements from jQuery Mobile.

Responsive Design

A jQuery Mobile UI will render responsively across different display sizes. For example, the same UI will display appropriately on phones (see Figure 1–9) or larger devices such as tablets, desktops, or TVs (see Figure 1–10).

Figure 1–9. *Phone Display*

Figure 1–10. *Tablet/Desktop/TV Display*

The Build Once, Run Anywhere Myth

Is it possible to build a single application that is universally available for all consumers (phones, desktops, tablets)? Yes, it is possible. The Web provides universal distribution. jQuery Mobile provides cross-browser support. And with CSS media queries we can begin tailoring our UI to best fit the form factor. For example, on small devices we can serve small images with brief content whereas on larger devices we may serve up larger images with detailed content. Today, most organizations with a mobile presence typically support both a desktop Web and a mobile site. There is waste any time you must support multiple distributions of an application. The rate at which organizations are embracing mobile presences, combined with their need to avoid waste, will drive the build once run anywhere myth to fruition.

Responsive Forms

In certain situations, jQuery Mobile will create responsive designs for you. The following figures show how well jQuery Mobile's responsive design applies to form field positioning in portrait versus landscape mode. For instance, in the portrait view (see Figure 1–11) the labels are positioned above the form fields. Alternatively, when repositioning the device in landscape (see Figure 1–12) the form fields and labels appear side-by-side. This responsive design provides the most usable experience based on the devices available to screen real estate. jQuery Mobile provides many of these good UX principles for you without any effort on your part!

Figure 1–11. *Responsive Design (portrait)*   **Figure 1–12.** *Responsive Design (landscape)*

Themable Styling

jQuery Mobile supports a themable design that allows designers to quickly re-style their UI. By default, jQuery Mobile provides five themable designs with the flexibility to interchange themes for all components including page, header, content, and footer components. The most useful tool for creating custom themes is ThemeRoller[3].

Restyling a UI takes minimal effort. For example, I can quickly take a default themed jQuery Mobile application (see Figure 1–13) and re-style it with another built-in theme in seconds. In the case of my modified theme (see Figure 1–14), I chose an alternate theme from the list. The only markup required was an addition of a data-theme attribute. We will discuss themes in greater detail in Chapter 7.

```
<!-- Set the lists background to black -->
<ul data-role="listview" data-inset="true" data-theme="a">
```

---

[3] See http://jqueryui.com/themeroller/. ThemeRoller is a web-based tool that automates the process of generating new CSS-based themes.

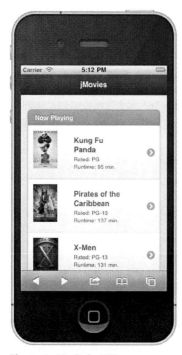

Figure 1–13. *Default Theme*   **Figure 1–14.** *Alternate Theme*

Accessible

jQuery Mobile apps are 508 compliant by default, a characteristic that is valuable to anyone.[4]

In particular, government or state agencies require their applications to be 100% accessible. Furthermore, mobile screen reader usage is rising. According to WebAIM,[5] 66.7% of screen reader users use the screen reader on their mobile device.

---

[4] 508 Compliance is a federal law that requires applications to be accessible by users with disabilities. The most commonly used assistive technologies on the mobile Web are screen readers.

[5] See http://webaim.org/projects/screenreadersurvey3/#mobileusage.

> **TIP:** Interested in testing your mobile site for 508 compliance? Evaluate your mobile site with WAVE.[6]

In addition to testing your mobile apps accessibility with WAVE it is also valuable to physically test your Mobile Web application with an actual assistive technology. For example, if you have an iOS device, activate Apple's Accessibility tool, VoiceOver[7] and experience the behavior first hand.

> **NOTE:** If you are interested in viewing existing jQuery Mobile applications, an online jQuery Mobile Gallery is available for ideas and inspiration (see `http://www.jqmgallery.com/`).

Summary

In this chapter, we reviewed the important features that make jQuery Mobile unique:

- jQuery Mobile apps are universally available to all devices with a browser and are optimized to run on nearly every phone, tablet, desktop, and e-reader device available today.

- jQuery Mobile applications can take advantage of the instant deployment capabilities we have grown accustomed to on the Web.

- A single jQuery Mobile code base will render consistently across all supported platforms without customizations. This is a very cost-effective solution when compared to the alternative of building an app for each OS or client.

- jQuery Mobile is a simplified markup-driven framework that should appear very familiar to Web designers and developers. You may be very surprised and excited by the fact that you can build jQuery Mobile apps with 100% markup!

- jQuery Mobile utilizes progressive enhancement techniques to render a very rich experience for all A-grade devices and provides a usable experience for older C-grade browsers.

- A jQuery Mobile UI will render responsively across devices of various sizes including phones, tablets, desktops, or TV's.

- jQuery Mobile supports a themable design that allows designers to quickly re-style their UI globally.

- All jQuery Mobile applications are 508 compliant.

---

[6] See `http://wave.webaim.org/`.

[7] See `http://www.apple.com/accessibility/iphone/vision.html`.

Chapter 2

Getting Started with jQuery Mobile

In Chapter 1, we reviewed the characteristics that make jQuery Mobile unique. Now we are going to review the basics of jQuery Mobile so we can get up and running quickly. We will start with an overview of the jQuery Mobile page template. There are actually two page templates you may choose from and we will discuss the advantages of each. Next, we will peek under the hood and see how jQuery Mobile enhances our semantic markup into an optimized mobile experience. Also, we will explore how the jQuery Mobile navigation model works. Although jQuery Mobile manages the entire navigational experience it is important to have an understanding of how the navigation model works. And lastly, we will show you how to make your page transitions really "pop." Anxious to get rolling? Let's begin with an example of a jQuery Mobile page.

jQuery Mobile Page Template

A jQuery Mobile page template is shown in Listing 2–1. Before we continue any further, let's get up and running. Copy the HTML template (ch2/template.html), paste it on your desktop and launch it from your favorite browser. You are now running a jQuery Mobile app that should look identical to Figure 2–1 regardless of what browser you are using! The template is semantic HTML5 and contains the jQuery Mobile specific attributes and asset files (CSS, js). Each specific jQuery Mobile asset and attribute is highlighted and explained in Listing 2–1.

Listing 2–1. *jQuery Mobile Page Template (ch2/template.html)*

```
<!DOCTYPE html>
<html>
<head>
    <meta charset="utf-8">
    <title>Title</title>
    <meta name="viewport" content="width=device-width, initial-scale=1">      1
    <link rel="stylesheet" type="text/css" href="jquery.mobile.css" />        2
    <script type="text/javascript" src="jquery.js"></script>                  3
```

```
        <!--<script src="custom-scripts-here.js"></script>-->            4
        <script type="text/javascript" src="jquery.mobile.js"></script>   5
</head>
<body>

<div data-role="page">                                                   6
    <div data-role="header">                                             7
        <h1>Page Header</h1>
    </div>

    <div data-role="content">                                            8
        <p>Hello jQuery Mobile!</p>
    </div>

    <div data-role="footer">                                             9
        <h4>Page Footer</h4>
    </div>
</div>

</body>
</html>
```

1. This is the recommended viewport configuration for jQuery Mobile. The device-width value indicates we want the content to scale the full width of the device. The initial-scale setting sets the initial scaling or zoom factor used for viewing a Web page. A value of 1 displays an unscaled document. As a jQuery Mobile developer you can customize the viewport settings to your application needs. For example, if you wanted to disable zoom you would add user-scalable=no. However, disabling zoom is a practice you would want to do sparingly because it breaks accessibility.

2. jQuery Mobile's CSS will apply stylistic enhancements for all A-Grade and B-Grade browsers. You may customize or add your own CSS as necessary.

3. The jQuery library is a core dependency of jQuery Mobile and it is highly recommended to leverage jQuery's core API within your Mobile pages too if your app requires more dynamic behavior.

4. If you need to override jQuery Mobile's default configuration you can apply your customizations here. Refer to Chapter 8, Configuring jQuery Mobile, for details on customizing jQuery Mobile's default configuration.

5. The jQuery Mobile JavaScript library must be declared after jQuery and any custom scripts you may have. The jQuery Mobile library is the heart that enhances the entire mobile experience.

6. data-role="page" defines the page container for a jQuery Mobile page. This element is only required when building multi-page designs (see Listing 2–3).

7. data-role="header" is the header or title bar as shown in Figure 2–1. This attribute is optional.

8. `data-role="content"` is the wrapping container for the content body. This attribute is optional.

9. `data-role="footer"` contains the footer bar as shown in Figure 2–1. This attribute is optional.

> **IMPORTANT:** The sequence of the CSS and JavaScript files must appear in the order as listed in Listing 2–1. The ordering is necessary to properly initialize the dependencies before they are referenced by jQuery Mobile. Additionally, it is recommended to download these files from a Content Delivery Network (CDN). In particular, you may download them from the jQuery Mobile CDN.[1] The files from the CDNs are highly optimized and will provide a more responsive experience for your users. They are compressed, cached, minified, and can be loaded in parallel!

Figure 2–1. *Hello jQuery Mobile*

---

[1] See `http://jquerymobile.com/download/`.

> **TIP:** To position the footer at the very bottom of the screen, add `data-position=`
> `"fixed"` to the footer element. A default footer is positioned after the content and not at the
> bottom of the device. For instance, if your content only consumed half the device height, the
> footer would appear in the middle of the screen.
>
> `<div data-role="footer" `**`data-position="fixed">`**

jQuery Mobile Page Enhancements

How does jQuery Mobile enhance the markup for an optimized mobile experience? For a
visual, refer to Figure 2–2.

1. First, jQuery Mobile will load the semantic HTML markup (see Listing 2–1).

2. Next, jQuery Mobile will iterate each page component, defined by their data-role
   attribute. As jQuery Mobile iterates each page component it will enhance the
   markup and apply mobile optimized CSS3 enhancements for each component.
   jQuery Mobile ultimately enhances the markup into a page that renders universally
   across all mobile platforms.

3. Lastly, after the page enhancements are complete, jQuery Mobile will show the
   optimized page. See Listing 2–2 for a view of the enhanced source that gets
   rendered by the mobile browser.

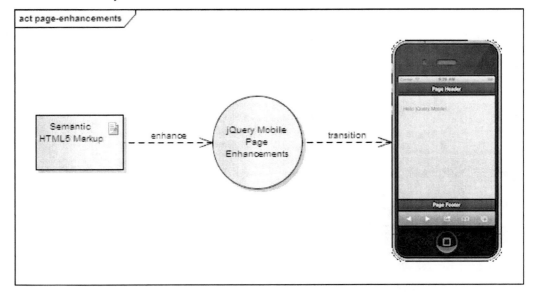

Figure 2–2. *jQuery Mobile Page Enhancements Diagram*

Listing 2–2. *jQuery Mobile Enhanced DOM*

```html
<!DOCTYPE html>
<html class="ui-mobile">
 <head>
   <base href="http://www.server.com/app-name/path/">                           1
   <meta charset="utf-8">
   <title>Page Header</title>
   <meta content="width=device-width, initial-scale=1" name="viewport">
   <link rel="stylesheet" type="text/css" href="jquery.mobile-min.css" />
    <script type="text/javascript" src="jquery-min.js"></script>
    <script type="text/javascript" src="jquery.mobile-min.js"></script>
 </head>

 <body class="ui-mobile-viewport">                                             2
   <div class="ui-page ui-body-c ui-page-active" data-role="page"
        style="min-height: 320px;">
     <div class="ui-bar-a ui-header" data-role="header" role="banner">
       <h1 class="ui-title" tabindex="0" role="heading" aria-level="1">
          Page Header
       </h1>
     </div>

     <div class="ui-content" data-role="content" role="main">
        <p>Hello jQuery Mobile!</p>
     </div>

     <div class="ui-bar-a ui-footer ui-footer-fixed fade ui-fixed-inline"
          data-position="fixed" data-role="footer" role="contentinfo"
          style="top: 508px;">
       <h4 class="ui-title" tabindex="0" role="heading" aria-level="1">
          Page Footer
       </h4>
     </div>
   </div>

   <div class="ui-loader ui-body-a ui-corner-all" style="top: 334.5px;">
     <span class="ui-icon ui-icon-loading spin"></span>
     <h1>loading</h1>
   </div>

 </body>
</html>
```

1. The base tag's @href specifies a default address or the default target for all links
    on a page. For example, jQuery Mobile will leverage @href when loading page-
    specific assets (images, CSS, js, etc.).

2. The body tag contains the enhanced styling for the header, content, and footer
    components. By default, all components have been styled with the default theme
    and their mobile-specific CSS enhancements. As an added bonus, all
    components now support accessibility thanks to the addition of the WAI-ARIA
    roles and levels. You get all these enhancements for free!

By now you should feel comfortable designing a basic jQuery Mobile page. You have been introduced to the core page components (page, header, content, footer) and have seen the resulting DOM of an enhanced jQuery Mobile page. Next we will explore jQuery Mobile's multi-page template.

Multi-Page Template

jQuery Mobile supports the ability to embed multiple pages within a single HTML document as shown in Listing 2–3. This strategy can be used to prefetch multiple pages up front and achieve quicker response times when loading sub-pages. As you can see in the example below, the multi-page document is identical to the single-page template we saw earlier except a second page has been appended after the first page. The multi-page specific details are highlighted and discussed below.

Listing 2–3. *Multi-Page Template (ch2/multi-page.html)*

```
<head>
  <meta charset="utf-8">
  <title>Multi Page Example</title>
  <meta name="viewport" content="width=device-width, initial-scale=1">
  <link rel="stylesheet" type="text/css" href="jquery.mobile-min.css" />
  <script type="text/javascript" src="jquery-min.js"></script>
  <script type="text/javascript">
    /* Shared scripts for all internal and ajax-loaded pages */
  </script>
  <script type="text/javascript" src="jquery.mobile-min.js"></script>
</head>

<body>

<!-- First Page -->
<div data-role="page" id="home" data-title="Home">                    1
    <div data-role="header">
        <h1>Welcome Home</h1>
    </div>

    <div data-role="content">
        <a href="#contact" data-role="button">Contact Us</a>        2
    </div>
</div>

<!-- Second Page -->
<div data-role="page" id="contact" data-title="Contact">
    <div data-role="header">
        <h1>Contact Us</h1>
    </div>

    <div data-role="content">
        Contact information...
    </div>
    <script type="text/javascript">
        /* Page specific scripts here */                            3
    </script>
</div>

</body>
```

1. Each page in a multi-page document must contain a unique `id`. A page can have a `data-role` of either `page` or `dialog`. When the multi-page document is initially shown only the first page is enhanced and displayed. For example, when the `multi-page.html` document is requested the page with `id="home"` will be shown because it is the first page listed in the multi-page document. If you wanted to request the page with `id="contact"` you would request the multi-page document with the hash of the internal page you wanted displayed (`multi-page.html#contact`). When a multi-page document loads only the initial page will be enhanced and shown. Subsequent pages will be enhanced as they get requested and cached within the DOM. This behavior is ideal for quick response times. To set the title for each internal page, add the `data-title` attribute.

2. When linking to an internal page you must refer to it by page `id`. For example, the href to link to the contact page must be set as `href="#contact"`.

3. If you want to scope scripts to a specific page they must be placed within the page container. This rule also applies to pages that get loaded via Ajax and we will discuss this further in the next section. For example, any JavaScript declared internally on `multi-page.html#contact` will not be accessible to `multi-page.html#home`. Only the scripts of the active page will be accessible. However, all scripts, including jQuery, jQuery Mobile, and your own custom scripts declared within the parent document's head tag will be available to all internal and Ajax-loaded pages.

Setting the Page Title of an Internal Page

It's important to note that the title of an internal page will be set according to the following order of precedence:

1. If a `data-ti2-tle` value exists, it will be used as the title for the internal page. For example, the title for `"multi-page.html#home"` will be set to `"Home"`.

2. If no `data-title` value exists, the header will be used as the title for the internal page. For example, if no `data-title` attribute existed for `"multi-page.html#home"`, the title would be set to `"Welcome Home"`, the value of its header tag.

3. Lastly, if neither a `data-title` nor header exists on the internal page the title element within the head tag will be used as the title for the internal page. For example, if no `data-title` attribute and no header existed for `"multi-page.html#home"`, the title would be set to `"Multi Page Example"`, the value of the parent document's title tag.

IMPORTANT: When linking to a page that contains multiple pages, you must add
`rel="external"` to its link.

```
<!-- Must include rel="external" when linking to multi-page documents -->
<a href="multi-page.html" rel="external">Home</a>

<!-- May optionally use the target attribute -->
<a href="multi-page.html" target="_blank">Home</a>
```

This will perform a full-page refresh. It is required because jQuery Mobile cannot load a multi-page document into the DOM of an active page. It would cause a namespace collision with how jQuery Mobile leverages the URL hash (#). jQuery Mobile leverages the hash value to identify internal pages within a multi-page document.

Additionally, because jQuery Mobile leverages the hash to identify unique pages within the DOM it is not possible to leverage the anchor tag bookmark feature (`index.html#my-bookmark`). jQuery Mobile treats `my-bookmark` as a page identifier, not a bookmark. Ajax-Driven Navigation will be discussed in greater detail in the next section.

Single-Page versus Multi-Page Documents

You will need to identify page access trends to determine which strategy makes most sense from a bandwidth and response time perspective. Multi-page documents will consume more bandwidth on their initial load but they only require a single server request and as a result their sub-pages are loaded with very fast response times. A single-page document will consume less bandwidth per request but they require a server request per page which results in much slower response times.

If you have several pages that are commonly accessed in sequence they make an ideal candidate to load upfront within the same document. The initial bandwidth hit is slightly higher but we achieve instant responses when accessing the next page. However, if the probability is low that the user will access both pages then you should opt to keep the files separate and achieve a lower bandwidth hit on the initial load.

There are tools available to help you collect page access trends and other metrics to help optimize your page access strategy. For example, Google Analytics[2] or Omniture[3] are common analytic solutions for Mobile Web applications.

---

[2] See http://www.google.com/analytics/.

[3] See http://www.omniture.com/.

> **TIP:** In most cases it is recommended to leverage the single-page model and dynamically append popular pages to the DOM in the background. We can achieve this behavior by adding the `data-prefetch` attribute to any link we want to dynamically load:
>
> ```
> <a href="load-page-dynamically.html" data-prefetch></a>
> ```
>
> This hybrid approach allows us to *selectively* choose which links we want to load and cache. Again, this pattern is only recommended for pages that are accessed very frequently because this behavior will trigger an additional HTTP request to load the page dynamically.

Ajax-Driven Navigation

In the multi-page example above (see Listing 2–3) we saw how jQuery Mobile navigates from one internal page to another. When the multi-page document was initialized, both internal pages were already added to the DOM so the page transition from one internal page to the other was extremely fast. When navigating from page to page we can configure the type of transition to apply. By default, the framework will apply a "slide" effect for all transitions. We will discuss transitions and the types of transitions we can choose from later in the chapter.

```
<!-- Navigate to an internal page -->
<div data-role="content">
    <a href="#contact" data-role="button">Contact Us</a>
</div>
```

The navigation model is different when a single-page transitions to another single-page. For instance, we could extract the contact page from our multi-page into its own file (contact.html). Now our home page (hijax.html) can access the contact page as a normal HTTP link reference:

Listing 2–4. *Ajax Navigation (ch2/hijax.html)*

```
<div data-role="content">
    <a href="contact.html" data-role="button">Contact Us</a>
</div>
```

When clicking the "Contact Us" link above, jQuery Mobile will process that request as follows:

1. jQuery Mobile will parse the `href` and load the page via an Ajax request (Hijax). For a visual, refer to Figure 2–3. If the page is loaded successfully, it will be added to the DOM of the current page.

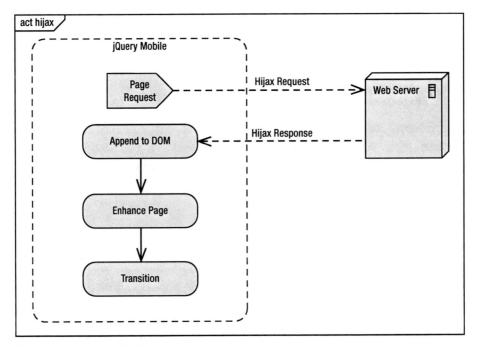

Figure 2–3. *jQuery Mobile Hijax Request*

With the page successfully added to the DOM, jQuery Mobile will enhance the page as necessary, update the base element's @href, and set the data-url attribute if it was not explicitly set.

2. The framework will then transition to the new page with the default "slide" transition applied. *The framework is able to achieve a seamless CSS transition because both the "from" and "to" pages exist together in the DOM.* After the transition is complete, the page that is currently visible or active will be assigned the "ui-page-active" CSS class.

3. The resulting URL is also bookmarkable. For example, if you want to deep link to the contact page you may access it from its full path: http://<host:port>/ch2/contact.html.

> **NOTE:** As an added bonus, Ajax-based navigation will also produce clean URLs in browsers that support HTML5's pushState. This feature is supported in the recent versions of desktop Safari, Chrome, Firefox, and Opera. Android (2.2+) and iOS5 also support pushState. In browsers that do not support this feature, the hash-based URLs (http://<host:port>/ hijax.html#contact.html) will be used to preserve the ability to share and bookmark URLs.

4. If any page fails to load within jQuery Mobile, a small error message overlay of "Error Loading Page" will be shown and faded out (see Figure 2–4).

Figure 2–4. *Error Loading Screen*

$.mobile.changePage()

- The changePage function handles all the details of transitioning from one page to another. You are allowed to transition to any page except the same page. The complete list of available transition types are shown in Table 2–1.

Usage

- $.mobile.changePage(toPage, [options])

Arguments

- **toPage** (string or jQuery collection). The page to transition to.

  - toPage (string). A file URL (`"contact.html"`) or internal element's ID (`"#contact"`).

  - toPage (jQuery collection). A jQuery collection containing a page element as its first argument.

- **options** (object). A set of key/value pairs that configure the changePage request. All settings are optional.

  - **transition** (string, default: `$.mobile.defaultTransition`).The transition to apply for the change page. The default transition is `"slide"`.

  - **reverse** (boolean, default: `false`). To indicate if the transition should go forward or reverse. The default transition is forward.

  - **changeHash** (boolean, default: `true`). Update the hash to the page's URL when page change is complete.

  - **role** (string, default: `"page"`). The data-role value to be used when displaying the page. For dialogs use `"dialog"`.

  - **pageContainer** (jQuery collection, default: `$.mobile.pageContainer`). Specifies the element that should contain the page after it is loaded.

- **type** (string, default: "get"). Specifies the method ("get" or "post") to use when making a page request.

- **data** (string or object, default: `undefined`). The data to send to an Ajax page request.

- **reloadPage** (boolean, default: false). Force a reload of the page, even if it is already in the DOM of the page container.

- **showLoadMsg** (boolean, default: true). Display the loading message when a page is requested.

- **fromHashChange** (boolean, default: false). To indicate if the changePage came from a hashchange event.

Example #1:

```
//Transition to the "contact.html" page.
$.mobile.changePage( "contact.html" );

<!-- Markup equivalent -->
<a href="contact.html">Contact Us</a>
```

Example #2:

```
// Go to an internal "#contact" page with a reverse "pop" transition.
$.mobile.changePage( '#contact', { transition: "pop", reverse: true } );

<!-- Markup equivalent -->
<a href="contact.html" data-transition="pop" data-direction="reverse">
  Contact
</a>
```

Example #3:

```
/* Dynamically create a new page and open it */

// Create page markup
var newPage = $("<div data-role=page data-url=hi><div data-role=header>
  <h1>Hi</h1></div><div data-role=content>Hello Again!</div></div>");

// Add page to page container
newPage.appendTo( $.mobile.pageContainer );

// Enhance and open new page
$.mobile.changePage( newPage );
```

> **IMPORTANT:** Ajax navigation will not be used for situations where you load an external link:
>
> ```
> <!-- Ajax navigation will be ignored when loading a page with a
> rel="external" or target attribute -->
> <a href="multi-page.html" rel="external">Home</a>
> ```
>
> ```
> <!-- Ajax navigation will be ignored -->
> <a href="multi-page.html" target="_blank">Home</a>
> ```
>
> Under these conditions, normal HTTP request processing will occur. Furthermore, CSS transitions will not be applied. As mentioned earlier, the framework is able to achieve smooth transitions by dynamically loading the "from" and "to" pages into the same DOM and then applying a smooth CSS transition. Without Ajax navigation the transition will not appear as smooth and the default loading message ($.mobile.loadingMessage) will not be shown during the transition.

Configuring Ajax Navigation

Ajax navigation is enabled globally but you can disable this feature if DOM size is a concern or if you need to support a particular device that does not support hash history updates (see Note below). By default, jQuery Mobile will manage the DOM size or cache for us with only the "from" and "to" pages involved in the active page transition merged into the DOM. To disable Ajax navigation set $.mobile.ajaxEnabled = false when binding to the mobileinit event. For further information about configuring jQuery Mobile or managing the DOM cache refer to Chapter 8.

> **NOTE:** Ajax navigation has been disabled on platforms that have known conflicts with hash history updates. For instance, jQuery Mobile has disabled Ajax navigation ($.mobile.ajaxEnabled = false) for BlackBerry 5, Opera Mini (5.0-6.0), Nokia Symbian^3, and Windows Phone 6.5. These devices are more usable when browsing with regular HTTP and full-page refreshes.

Transitions

jQuery Mobile has six CSS-based transition effects to choose from when transitioning between pages. By default, the framework will apply a "slide" effect for all transitions. We can set an alternate transition by adding the data-transition attribute to any link, button, or form:

```
<a href="dialog.html" data-transition="slideup">Show Dialog</a>
```

The complete list of transition effects are described in Table 2–1:

Table 2–1. *Transition Effects*

Transition	Common Usage
slide	The most common transition for moving between pages. This transition gives the appearance of moving forward or backward in a flow of pages. This is the default transition for all links.
slideup	A common transition for opening dialogs or showing additional information. This transition gives the appearance that you need to collect additional input for the page that is currently active.
slidedown	This transition is the inverse of slideup but can be used for a similar effect.
pop	Another transition for opening dialogs or showing additional information. This transition gives the appearance that you need to collect additional input for the page that is currently active.
fade	A common transition effect for entry or exit pages.
flip	A common transition for showing additional information. Typically, the back of a screen displays configuration options (info icon) that do not need to be in the main UI.
none	No transition will be applied.

The process of transitioning from page-to-page occurs in the following steps:

1. A user taps the button to navigate to the next page (see Figure 2–5).

2. The framework will load the next page with a Hijax request and add it to the DOM of the current page. With both pages essentially side-by-side, a smooth transition is ready to occur (see Figure 2–6).

3. The framework transitions to the next page (see Figure 2–7). This example uses the default "slide" transition.

4. The next page is shown and the transition is complete (see Figure 2–8).

Figure 2–5. *Step #1: Tap the button to navigate to another page*

Figure 2–6. *Step #2: Framework loads the next page side-by-side*

Figure 2–7. *Step #3: Framework transitions to next page*

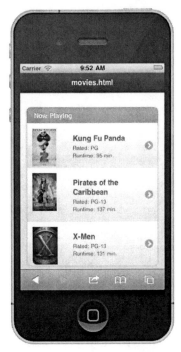

Figure 2–8. *Step #4: Transition is complete*

TIP: You can set a "backward" transition by adding `data-direction="reverse"` to your links. A forward `"slide"` transition will slide left and conversely a reverse `"slide"` transition will slide right. For instance, a reverse transition is applied by default when transitioning back in history. However, if you have a "home" link on your header you will need to apply the **data-direction="reverse"** attribute otherwise the default "forward" effect would occur:

```
<a href="home.html" data-icon="home" data-iconpos="notext"
    data-direction="reverse" class="ui-btn-right jqm-home">
    Home
</a>
```

Dialogs

Dialogs are similar to pages except their border is inset to give them the appearance of a modal dialog. jQuery Mobile is quite flexible in regards to how we can style our dialogs. We can create confirmation dialogs (see Figure 2–9), alert dialogs (see Figure 2–10), and even action sheet styled dialogs (see Figure 2–11, Figure 2–12).

Figure 2–9. *Confirmation Dialog (ch2/dialog.html)*   **Figure 2–10.** *Alert Dialog (ch2/alert.html)*

We can transform a page into a dialog at either the link or page component. On a link, add the `data-rel="dialog"` attribute as shown in Listing 2–5. The addition of this attribute will automatically load the target page and enhance it as a modal dialog.

Listing 2–5. *Link-level transformation*

```
<!-- Open a page as a dialog -->
<a href="#terms" data-rel="dialog" data-transition="slidedown">Terms</a>


<!-- The page remains unchanged. -->
<div data-role="page" id="terms">
 <div data-role="header">
  <h1>Terms and Conditions</h1>
 </div>

 <div data-role="content">
  Do you agree to these terms?


  <a href="#" data-role="button" data-inline="true"
     data-rel="back" data-theme="a">Disagree</a>
  <a href="#" data-role="button" data-inline="true">Agree</a>
 </div>
</div>
```

We can also configure dialogs at the page container. Add the `data-role="dialog"` attribute to the page container and when the component loads it will be enhanced as a modal dialog (see Listing 2–6).

Listing 2–6. *Page-level transformation (ch2/dialog.html)*

```
<!-- Link without data-rel="dialog" attribute -->
<a href="#terms" data-transition="slidedown">Terms and Conditions</a>


<!-- Configure this page to appear as a dialog -->
<div data-role="dialog" id="terms">
 <div data-role="header">
  <h1>Terms and Conditions</h1>
 </div>

 <div data-role="content" data-theme="c">
  Do you agree to these terms?

  <a href="#" data-role="button" data-inline="true"
     data-rel="back" data-theme="a">Disagree</a>
  <a href="#" data-role="button" data-inline="true">Agree</a>
 </div>
</div>
```

> **NOTE:** Any link with `data-rel="dialog"` or any page with `data-role="dialog"` will not
> appear in history and cannot be bookmarked. For example, if you navigate to a dialog, close the
> dialog and then tap the browser's forward button, you will not go forward to the dialog because it
> will not exist in history.

Link versus Page Configuration

With two options available for opening dialogs, which should we choose? I prefer the page configuration (data-role="dialog") because it allows us to configure the dialog once at the page container and the buttons that navigate to our dialog require zero modifications. For example, if we have three buttons linking to our dialog, the page-based configuration only requires one modification. Whereas a link-based configuration would require three changes, one to each button.

The jQuery Mobile dialog API also exposes a close method that you may leverage when programmatically working with dialogs. For example, if we wanted to programmatically handle the processing of the "Agree" button in Figure 2–9, we could handle the click event, process any necessary business logic, and close the dialog when complete:

```
function processAgreement(){
  // Save the agreement...

  // Close the dialog
  $('.ui-dialog').dialog('close');
}
```

Action Sheets

In addition to the traditional dialog we can also style our dialog as an action sheet (see Figures 2–11 and 2–12). Simply remove the header, add minor styling updates (see Listing 2–7) and your dialog appears as an action sheet. Action sheets are commonly used to solicit a response from the user. For the best user experience, it is recommended to use the slidedown transition for an action sheet. Conveniently, when dialogs close they automatically apply the reverse transition. For example, when you close this action sheet the reverse slideup transition will be applied.

Figure 2–11. *Action Sheet #1*
(ch2/action-sheet1.html)

Figure 2–12. *Action Sheet #2*
(ch2/action-sheet2.html)

Listing 2–7. *Action Sheet (ch2/action-sheet1.html)*

```html
<!-- Logout link -->
<a href="#logout" data-transition="slidedown">Logout</a>


<!-- Create an action sheet by simply removing the header! -->
<div data-role="dialog" id="logout">
  <div data-role="content">
    <span class="title">Are you sure?</span>

    <a href="#home" data-role="button" data-theme="b">Yes, I'm Sure!</a>
    <a href="#" data-role="button" data-theme="c" data-rel="back">No Way!</a>
  </div>
  <style>
    span.title { display:block; text-align:center;
                 margin-top:10px; margin-bottom:20px; }
  </style>
</div>
```

This is also our first exposure to the data-theme attribute. We can simply add contrast and style to all jQuery Mobile components with this attribute. In our dialog examples, we can set the theme of our background and buttons. When styling dialog buttons it is common to contrast the style of the cancel and action buttons. Themes within jQuery Mobile are discussed in greater depth in Chapter 7.

Dialog UX Guidelines

Consistency is the most important design goal when styling your UI components. In regards, to dialog-specific guidelines a few tips from Apple's Mobile Interface Guidelines[4] include:

> **TIP:** Dialogs have a their maximum width set to 500 pixels by default. This will appear full screen on smaller mobile displays and will appear 500 pixels wide on desktop and tablet screens. If you need to override the default width use the following CSS in your theme:
>
> ```
> .ui-dialog .ui-header, .ui-dialog .ui-content, .ui-dialog .ui-footer {
> max-width: 100%; }
> ```

Alerts:

- Prefer alerts to display important information that affects the use of the application (see Figure 2–10). An alert is *not user initiated*.

- Alert buttons are either colored light or dark. For a single-button alert the button is always light-colored. For a two-button dialog, the left button is always dark and the right button is always light (see Figure 2–9).

- In a two-button dialog that proposes a favorable action that people are likely to choose, the button that cancels the action should be on the left and dark-colored (see Figure 2–9).

- In a two-button dialog that proposes a potentially risky action (delete), the button that cancels the action should be on the right and light-colored. Often buttons that perform risky actions are red.

Action Sheets:

- Prefer action sheets to gather confirmation of *user-initiated* tasks (see Figure 2–11). Action sheets may also be used to provide the user a range of choices for their current task (see Figure 2–12).

- An action sheet always contains at least two buttons that allow the user to choose how to complete their task.

---

[4] See http://developer.apple.com/library/ios/documentation/userexperience/conceptual/mobilehig/MobileHIG.pdf.

    ▨   Include a cancel button to allow users to abandon the task. The cancel button is placed at the bottom of the action sheet to encourage users to read through all options before making a choice. The cancel button color should correspond to the color of the background.

Responsive Layouts with Media Queries

To create responsive designs with jQuery Mobile it is recommended to leverage the power of CSS3 Media Queries.[5] For example, if you need to enhance your layout for a particular device orientation you can detect the orientation of the device with media queries and apply your CSS modifications as necessary:

```
@media (orientation: portrait) {
  /* Apply portrait orientation enhancements here... */
}

@media (orientation: landscape) {
  /* apply landscape orientation enhancements here... */
}
```

In certain situations, jQuery Mobile will create responsive designs for you. The following figures show how well jQuery Mobile's responsive design behaves with form field positioning in portrait versus landscape mode. For instance, in the portrait view (see Figure 2–13) the labels are positioned above the form fields. Alternatively, when repositioning the device in landscape (see Figure 2–14) the form fields and labels appear side-by-side. This responsive design provides the most usable experience based on the devices available screen real estate. jQuery Mobile provides many of these good UX principles for you!

---

[5] See http://www.w3.org/TR/css3-mediaqueries/.

Figure 2–13. *Responsive (portrait)*   **Figure 2–14.** *Responsive (landscape)*

> **WARNING:** If you launch Figure 2–14 (ch2/responsive.html) in iOS and switch to landscape you may have noticed there is an iOS scaling issue in Mobile Safari.[6] "When the meta viewport tag is set to content="width=device-width, initial-scale=1", or any value that allows user-scaling, changing the device to landscape orientation causes the page to scale larger than 1.0. As a result, a portion of the page is cropped off the right, and the user must double-tap (sometimes more than once) to get the page to zoom properly into view."
>
> Until this issue is resolved in Mobile Safari you have several options to remedy the issue:
>
> - You can disable zoom. Although, disabling zoom is a practice you will want to do sparingly because it breaks accessibility.
>
>     ```
>     <meta name="viewport" content="width=device-width,
>         minimum-scale=1.0, maximum-scale=1.0">
>     ```
>
> - You can dynamically adjust the meta tag when the user zooms.[7]

---

[6] See http://filamentgroup.com/examples/iosScaleBug/.

[7] See http://adactio.com/journal/4470/.

In the examples above (see Figure 2–13) jQuery Mobile is able to apply responsive designs by leveraging the min-max width media features. For example, form elements float beside their labels when the browser supports a width greater than 450 pixels. The CSS to support this behavior for text inputs looks like this:

```
label.ui-input-text {
    display: block;
}

@media all and (min-width: 450px){
  label.ui-input-text { display: inline-block; }
}
```

> **IMPORTANT:** Windows Phone 7 (Internet Explorer 8 and below) does not support media queries. If you would like to support responsive designs on browsers that do not support media queries it is recommended to leverage Respond.js.[8] Respond.js provides media query support for browsers that do not support them.

There is also a limited set of Webkit-specific media extensions you may find useful. For example, to apply CSS enhancements for newer iOS devices with a high-definition retina display you may use the `webkit-min-device-pixel-ratio` media feature:

```
//Webkit-specific media query for the iOS high-resolution Retina display @media screen
and (-webkit-min-device-pixel-ratio: 2){
   // Apply retina display enhancements
}
```

As an added bonus for the iOS users, jQuery Mobile has included a full set of retina-optimized icons that are automatically applied to any iOS device with a very high-resolution display.

> **NOTE:** If you choose to segregate your media-specific styles in separate files you can reference them with the HTML <link> media attribute. This practice promotes good separation of concerns but suffers from a performance perspective because each separate file requires an additional HTTP request:
>
> ```
> <link href="default.css" />
> <link media="all and (min-width:450px)" href="widescreen.css" />
> ```

---

[8] See https://github.com/scottjehl/Respond.

Summary

In this chapter, we reviewed the basics of jQuery Mobile and saw how quickly we can get up and running with a jQuery Mobile application. We reviewed both jQuery Mobile page templates and discussed the advantages of each in regards to performance and navigation flow. We also took a peek under the hood to see how jQuery Mobile enhances our semantic markup into an optimized mobile experience. Additionally, we reviewed all available page transitions and discussed common usage patterns for each. Lastly, we saw the many different ways of styling dialogs to create an effective interface for informing or gathering feedback from our users. In Chapter 3 we will take a closer look at navigating with jQuery Mobile and how we can best utilize our header and footer controls to manage our mobile applications data.

Navigating with Headers, Toolbars, and Tab Bars

All mobile applications need toolbars to help navigate or manage data on the screen. In this chapter we will review the jQuery Mobile components that provide these features. The main components are headers and footers. Headers are commonly used to display the page title and may optionally include controls to help navigate or manage objects on the screen. Footers are designed similarly to headers but their responsibility is typically managed with a toolbar or tab bar. Additionally, we will discover the capabilities of a segmented control. A segmented control is a specialized control we may position in our header or footer to help display alternate views of data. We will explore each of these components and demonstrate how you can style them with text, standard icons, and even customized icons.

Header Bar

The header bar displays the title of the current screen. Additionally, you can add buttons for navigation or add controls that manage items in the page. Although the header is optional, it is commonly used to provide a title for the active page. Let's begin by reviewing the header structure and look at how we can add additional controls to the header to help manage items on the page.

Header Basics

There are a few points of importance about the header. They include:

- The header is defined with the `data-role="header"` attribute.

- The header is an optional component.

- The back button will not be shown in the header unless you explicitly enable it. The back button is discussed in detail in the next section.

- You may adjust the theme of the header with the `data-theme` attribute. If no theme is set for the header it will inherit the theme from the page component. The default theme is black (`data-theme="a"`).

- All heading levels (H1-H6) are styled identically by default to maintain visual consistency.

- You can make the header fixed with the addition of the `data-position="fixed"` attribute.

> **TIP:** You may also use the header as a segmented control as shown in Figure 3–5. A segmented control allows the user to display different views of related data.

Header Structure

The basic usage of a header is to simply display the title of the active page. A header in its simplest form is shown below.

```
<div data-role="header">
  <h1>Header Title</h1>
</div>
```

Header Positioning

There are three styles available for positioning the header. They include:

- Default: A default header will be shown at the top edge of the screen and will slide out of view when you scroll.

  ```
  <div data-role="header">
    <h1>Default Header</h1>
  </div>
  ```

- Fixed: A fixed header will always remain positioned and visible at the top edge of the screen. However, during a scroll event the header will disappear until the scroll is finished. We can create a fixed header with the addition of the `data-position="fixed"` attribute.

  ```
  <div data-role="header" data-position="fixed">
    <h1>Fixed Header</h1>
  </div>
  ```

> **NOTE:** In order to achieve true fixed toolbars, a browser needs to either support position:fixed or overflow:auto. Fortunately, new releases of WebKit (iOS5) are beginning to support this behavior. In jQuery Mobile, we can enable this behavior by setting the `touchOverflowEnabled` configuration option to true (see "Configurable jQuery Mobile Options" in Chapter 8 for more details).

- Responsive: When we create a fullscreen page the contents will appear edge-to-edge and the header and footer will responsively appear and disappear based on a touch response. Fullscreen mode is a useful scenario for photo or video displays. To create a fullscreen page add `data-fullscreen="true"` to the page container and include the `data-position="fixed"` attribute on the header and footer elements (see Listing 3–1). For instance, in Figure 3–1 we have a fullscreen page that displays a photo. If a user taps the screen, the header and footer will responsively appear and disappear (see Figure 3–2). In this example, we have a photo viewer with the header showing the counter of our image deck and the footer displays a toolbar to help navigate, email, or delete images.

Listing 3–1. *Fullscreen (ch3/position-fullscreen.html)*

```
<div data-role="page" data-fullscreen="true">
  <div data-role="header" data-position="fixed">
    <h3>Header</h3>
  </div>

  <div data-role="content">
    <!-- Fullscreen content -->
  </div>

  <div data-role="footer" data-position="fixed">
    <h3>Footer</h3>
  </div>
```

Figure 3–1. *Fullscreen*

Figure 3–2. *Fullscreen with responsive header and footer*

NOTE: The browsers URL bar will be hidden in iOS and Android when viewing jQuery Mobile pages. This is a convenient feature that allows the user to view more available screen real estate and it smooths out the transitions. However, if you need to view the URL bar, drag the page down and the URL bar will become visible.

Header Buttons

There are situations when you will need to add controls to the header to help manage the screen contents. For example, save and cancel buttons are common controls that are available when editing data. There are three styles of buttons you may add to a header. They include:

- A button with only text.

- A button with only an icon (see Figure 3–4). An icon-only button requires the addition of two attributes: `data-icon` and `data-iconpos="notext"`. For the complete listing of `data-icon` values refer to Table 4-1.

- A button with text and an icon (see Figure 3–3). This button also requires the `data-icon` attribute. Examples of each are shown here:

```
<!-- A button with only text -->
<a href="#">Done</a>

<!-- A button with only an icon -->
<a href="#" data-icon="plus" data-iconpos="notext"></a>

<!-- A button with text and an icon -->
<a href="#" data-icon="check">Done</a>
```

Buttons with Text and Icons

In Figure 3–3, we have a header with a "Cancel" and a "Done" button to help manage the entry of a movie review. As shown in Listing 3–2, the button is styled as an ordinary link. We also attached an icon to each button with the `data-icon` attribute. To create a text-only button simply remove the `data-icon` attribute. Within a header, buttons are positioned according to their semantic order. For example, the first button will be left-aligned and the second button will be right-aligned. If your header only contains a single button you can right-align the button by adding `class="ui-btn-right"` to the button's markup.

Listing 3–2. *Header Buttons (ch3/header-buttons.html)*

```
<div data-role="header" data-position="inline">
  <a href="#" data-icon="delete">Cancel</a>
  <h1>Add Review</h1>
  <a href="#" data-icon="check">Done</a>
</div>
```

Figure 3–3. *Header with buttons*

Buttons with Only Icons

jQuery Mobile includes several standard icons (see Table 4-1) that you may use to create icon-only buttons. For instance, the "info" icon is commonly used with a "flip" transition to display configuration options or more information. The use of standard icons consumes very little real estate and their meaning is relatively consistent across all devices. For instance, if we wanted to add an item to an existing list we may choose to show a "plus" icon that allows users to add an entry to the list (see Figure 3–4). In this example we have a listing of movie reviews and users can tap the "add" icon to create their review. To create an icon-only button, two attributes are required as shown in Listing 3–3.

Listing 3–3. *Header with Icon (ch3/header-icons.html)*

```
<div data-role="header">
  <h1>Reviews</h1>
  <a href="#" data-icon="plus" data-iconpos="notext" class="ui-btn-right"></a>
</div>
```

Figure 3–4. *Header with icon*

Header bar with a segmented control

A segmented control is an inline set of controls that each display a different view. For example, the segmented control in Figure 3–5 shows movies by a specific category. This segmented control allows users to quickly view movies by the category of their choice: In Theatres, Coming Soon, or Top Rated.

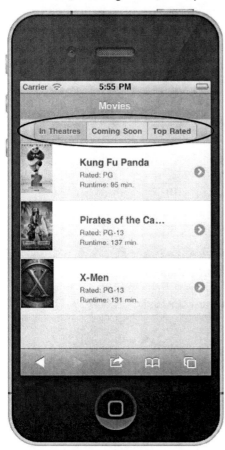

Figure 3–5. *Segmented Control*

It is recommended to position the segmented control within the main header as shown in Listing 3–4. This positioning allows the segmented control to integrate seamlessly with the main header if you choose to position the header as a fixed control. With the addition of a few minor styling updates we now have a segmented control that will allow users to quickly view data in alternate views!

Listing 3–4. *Segmented Control (ch3/header-segmented-control.html)*

```
<div data-role="header" data-theme="b" data-position="fixed">
  <h1>Movies</h1>
  <div class="segmented-control ui-bar-d">
```

```
            <div data-role="controlgroup" data-type="horizontal">
              <a href="#" data-role="button" class="ui-control-active">
                In Theatres
              </a>
              <a href="#" data-role="button" class="ui-control-inactive">
                Coming Soon
              </a>
              <a href="#" data-role="button" class="ui-control-inactive">
                Top Rated
              </a>
            </div>
          </div>
        </div>


        <style>
          .segmented-control { text-align:center;}
          .segmented-control .ui-controlgroup { margin: 0.2em; }
          .ui-control-active, .ui-control-inactive {
              border-style: solid; border-color: gray; }
          .ui-control-active { background: #BBB; }
          .ui-control-inactive { background: #DDD; }
        </style>
```

Fixing a Truncated Header or Footer...

jQuery Mobile will truncate headers and footers with long titles (see Figure 3–6). When the text is too long jQuery Mobile will truncate the text and add an ellipsis to the end. If you encounter this situation and want to show the complete text (see Figure 3–7) you can adjust the CSS selector to remedy the issue as shown in Listing 3–5.

Listing 3–5. *Truncation Fix (ch3/truncation-fixed.html)*

```
.ui-header .ui-title, .ui-footer .ui-title {
  margin-right: 0 !important; margin-left: 0 !important;
}
```

Figure 3–6. *Truncation Issue*

Figure 3–7. *Truncation Fix*

Back Button

Back buttons (see Figure 3–8) can generate great debates among UX designers. Should we add our own back buttons or should we leverage the hardware/software back buttons available on some devices and all browsers? Fortunately, jQuery Mobile offers you the choice of automatically enabling or disabling them globally. You also have the option of adding or removing them on a page-by-page basis.

Figure 3–8. *Back button must be explicitly enabled.*

The back button is disabled by default within jQuery Mobile. If you need the back button to appear within the header you have several options for adding them:

- You can add the back button to a specific page by adding `data-auto-back-btn="true"` on the page container.

- You can globally enable the back button by setting the `addBackBtn` option to `true` when binding to the `mobileinit` option. After setting this option, the back button will appear automatically if a page exists in the history stack. Under the covers, a back button simply executes `window.history.back()`. As shown below, you may also override the default back button text and theme. For instance, it is common to label back buttons with the title of the previous page. The `data-back-btn-text` attribute can be used for this convention. For additional details on setting global configuration options refer to Chapter 8, Configuring jQuery Mobile.

```
<!-- Show the back button and override the default back button text -->
<div data-role="page" data-add-back-btn="true"
    data-back-btn-text="Previous">
```

```
// Globally enable the back button, set the default back button text,
// and set back button theme
$(document).bind('mobileinit',function(){
    $.mobile.page.prototype.options.addBackBtn = true;
    $.mobile.page.prototype.options.backBtnText = "Previous";
    $.mobile.page.prototype.options.backBtnTheme = "b";
});
```

Furthermore, if you enabled the back button globally, you can choose to disable the back button on specific pages by adding the `data-add-back-btn="false"` attribute on the page header. This will remove the back button from the header of specific pages.

```
<!-- Disable the back button on a specific page if we globally enabled it -->
<div data-role="header" data-add-back-btn="false">
```

> **TIP:** Although back buttons are available in all mobile browsers there are a few specific cases within jQuery Mobile where you explicitly may need back buttons or alternative navigation:
>
> - It is recommended for all pages to include a link back to the home screen, either via a linked logo or home button. The goal is to never leave the user at a dead end within their navigation flow. A common scenario may arise when a user accesses a deep link or bookmarked page. If your only navigation mechanism is the back button and the history stack is empty the automatic back button will not appear, leaving the user at a dead end. Therefore, it is a very good practice to include a home icon link on the right side of the header bar.
> - When designing for PhoneGap integration you will need to consider the use of back buttons if your target OS does not support hardware-based navigation like iOS or WebOS.

Back Linking

If you want to create a button that behaves similarly to a back button you can add `data-rel="back"` to any anchor element:

```
<a href="home.html" data-rel="back" data-role="button">Go Back</a>
```

With `data-rel="back"`, the link will mimic the back button, going back one history entry (`window.history.back()`) and ignoring the link's default href. For C-Grade browsers or browsers with no JavaScript support the `data-rel` will be ignored and the `href` attribute will be used as a fallback.

Footer bar

The footer component is nearly identical to the header with only minor differences. The main difference is the footer is more flexible in regards to the placement of its buttons. For example, when working with the header the first button was left-aligned and the second button was right-aligned. The footer positions its buttons inline and in sequential order from left to right. This flexibility allows us to style our footer as a toolbar or tab bar. We will see examples of both but first let's begin with the basics.

Footer Basics

There are a few points of importance about the footer. They include:

- The footer is defined with the `data-role="footer"` attribute.

- The footer positions its buttons inline and in sequential order from left to right. This allows for the flexibility to create toolbars and tab bars.

- The footer is an optional component.

- You may adjust the theme of the footer with the `data-theme` attribute. If no theme is set for the footer it will inherit the theme from the page component. The default theme is black (`data-theme="a"`).

- You can make the footer fixed with the addition of the `data-position="fixed"` attribute.

- All footer levels (H1-H6) are styled identically by default to maintain visual consistency.

Footer Structure

The footer in its simplest form is shown in the code below. The data-role="footer" is the only required attribute. Within the footer, you may include any semantic HTML. Footers are commonly used to contain toolbar and tab controls. A toolbar provides a set of actions users can take in the current context. And a tab bar gives users the ability to switch between different views within the application.

```
<div data-role="footer">
  <!-- Add footer text or buttons here -->
</div>
```

> **TIP:** To position the footer at the very bottom of the screen, add data-position="fixed" to the footer element. A default footer is positioned after the content and not at the bottom edge of the screen (see Figure 3–9). For instance, if your content only consumed half the screen height the footer would appear in the middle of the screen. We can position the footer at the bottom of the screen by adding data-position="fixed" to the footer element.
>
> ```
> <div data-role="footer" data-position="fixed">
> ```

Figure 3–9. *Default footer position*

Footer Positioning

The three styles of positioning for the header also apply to the footer. They include:

- Default: A default footer is positioned after the content section. For instance, if your content extends beyond the height of your viewport the footer will not be shown until you scroll to the end of the content.

```
<div data-role="footer">
  <!-- Default footer -->
</div>
```

- Fixed: A fixed footer, will always remain positioned and visible at the bottom edge of the screen. However, during a scroll event the footer will disappear until the scroll is finished. We can create a fixed footer with the addition of the data-position="fixed" attribute.

```
<div data-role="footer" data-position="fixed">
  <h3>Fixed Footer</h3>
</div>
```

- Responsive: When we create a fullscreen page the contents will appear edge-to-edge and the header and footer will responsively appear and disappear based on a touch response. Fullscreen mode is a useful scenario for photo or video displays. To create a fullscreen page add data-fullscreen="true" to the page container and include the data-position="fixed" attribute on the header and footer elements. For an example refer to Figure 3–1.

Footer Buttons

There are three styles of buttons you may add to a footer. They include:

- A button with only text. This style of button works well within a toolbar because a toolbar's appearance is not as large as a tab bar. A normal link within the footer will display as a text-only button:

    ```
    <a href="#">Sync</a>
    ```

- A button with only an icon. This style of button also works well within a toolbar. An icon-only button requires the addition of two attributes, data-icon and data-iconpos="notext":

    ```
    <a href="#" data-icon="plus" data-iconpos="notext"></a>
    ```

- For the complete listing of data-icon values refer to Table 4-1.

- A button with text and an icon. This style of button works well within a tab bar:

    ```
    <a href="#" data-icon="home">Home</a>
    ```

> **TIP:** jQuery Mobile is an excellent framework for building applications that display responsively across mobile, tablet, and desktop browsers. While the header and footer components provide a "native" feel on mobile devices they translate poorly when viewed on the desktop. If your jQuery Mobile application is targeted for a diverse set of browser sizes you may prefer to omit the header and footer components. As an alternative, you may find it more beneficial to add custom header or footer markup directly within the content section.

Toolbars

Toolbars help manage the contents of the current screen. For instance, mail apps often use toolbars to help manage your email. In situations where users need to perform actions related to objects on the current screen a toolbar provides a useful experience. When building a toolbar we have the option of using icons or text. In our examples below we will view toolbar examples that contain buttons styled with icons, text, and a segmented control.

Toolbar with Icons

Icon-only toolbars are most common. Their primary advantage is they consume less screen real estate when compared to a textual alternative. When selecting icons it is important to choose standard icons that express clear meaning. In Figure 3–10 we have a screen that displays a movie review. To help the user manage the review we also included a toolbar with standard icons. The toolbar allows the user to perform five

Figure 3–10. *Toolbar with standard icons possible actions:*

1. Navigate to the prior review.

2. Reply to the review with a comment.

3. Mark the review as a favorite.

4. Add a new review for the movie.

5. Navigate to the next review.

Creating the toolbar requires minimal markup (see Listing 3–6). We simply need an unordered list of buttons wrapped in a div with the `data-role="navbar"` attribute. The toolbar buttons are flexible and will be evenly spaced according to the width of the device. In this example we used icons from jQuery Mobile's standard suite of available icons (see Table 4-1).

Listing 3–6. *Toolbar (ch3/toolbar-icons-standard.html)*

```
<div data-role="footer" data-position="fixed">
  <div data-role="navbar">
    <ul>
      <li><a href="#" data-icon="arrow-l"></a></li>
      <li><a href="#" data-icon="back"></a></li>
      <li><a href="#" data-icon="star"></a></li>
      <li><a href="#" data-icon="plus"></a></li>
      <li><a href="#" data-icon="arrow-r"></a></li>
    </ul>
  </div>
</div>
```

> **TIP:** The navbar component works equally well with custom icons if you want to style your navbars with added flare! If interested in the custom icon solution, we will demonstrate that solution in Listing 3–10.

Toolbar with a Segmented Control

You can also put a segmented control in a toolbar to give users access to different perspectives of your application's data or to different application views. In Figure 3–11 we positioned our segmented control within the toolbar to allow users to display different views of their calendar data. As you may have noticed, this segmented control (see Listing 3–7) is identical to the segmented control example shown in our header example. We can reuse the segmented control across both our header and footer components. The segmented control is simply a set of buttons wrapped within a control group and styled according to your needs. In the next section we will show yet another usage of the segmented control with a tab bar!

Figure 3–11. *Toolbar with Segmented Control*

Listing 3–7. *Toolbar with Segmented Control (ch3/toolbar-segmented-control.html)*

```
<!-- Toolbar with a segmented control -->
<div data-role="footer" data-position="fixed" data-theme="d"
     class="segmented-control">
  <div data-role="controlgroup" data-type="horizontal">
    <a href="#" data-role="button" class="ui-control-active">List</a>
    <a href="#" data-role="button" class="ui-control-inactive">Day</a>
    <a href="#" data-role="button" class="ui-control-inactive">Month</a>
  </div>
</div>

<style>
  .segmented-control { text-align:center; }
  .segmented-control .ui-controlgroup { margin: 0.2em; }
  .ui-control-active, .ui-control-inactive { border-style: solid;
      border-color: gray; }
  .ui-control-active { background: #BBB; }
  .ui-control-inactive { background: #DDD; }
</style>
```

Tab Bars

We can also style our footer as a tab bar. A tab bar gives users the ability to switch between different views within the application. If you are not too familiar with tab bars their behavior is very similar to tab-based navigation you find on the Web. Tab bars are commonly positioned as a persistent footer at the bottom edge of the screen that remains accessible from every location in the application. Tab bars typically contain buttons that display both an icon and text for clarity. In the examples below we will look at three styles of tab bars. The first tab bar example will include several of the standard icons that are already available within jQuery Mobile. Secondly, we will see a tab bar example that uses custom icons. jQuery Mobile conveniently allows integration with custom icons of your choice. And lastly, we will combine our tab bar and our segmented control within the same UI to allow the user to navigate and view alternate forms of data from the same screen.

Tab Bar with Standard Icons

The simplest tab bar solution (see Figure 3–12) is one that uses jQuery Mobile's standard icon set, as detailed in Listing 3–8. For the complete listing of standard jQuery Mobile icons refer to Table 4-1. If you leverage these standard icons your tab bar will require no extra styling.

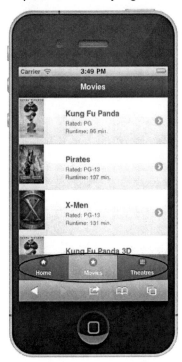

Figure 3–12. *Tab bar with standard Icons*

Listing 3–8. *Tab bar with standard icons* (*ch3/tabbar-icons-standard.html*)

```
<!-- tab bar with standard icons -->
<div data-role="footer" data-position="fixed">
  <div data-role="navbar">
   <ul>
    <li><a href="#" data-icon="home">Home</a></li>
    <li><a href="#" data-icon="star" class="ui-btn-active">
          Movies</a></li>
    <li><a href="#" data-icon="grid">Theatres</a></li>
   </ul>
  </div>
</div>
```

Persistent Tab Bar

In order to make our tab bars persistent we need to add an additional attribute to the footer. To keep the footer persistent during a page transition add the `data-id` attribute to the footer of each tab bar and set their values to the same identifier. For instance, in Listing 3–9, each tab bar contains an identifier of `data-id="main-tabbar"`. With this addition, your tab bar will remain persistent during a transition. For example, if we tapped on an inactive tab bar the screen would "slide" while the tab bar remains in a fixed and persistent state during the transition. Additionally, to retain the active state of each tab bar when transitioning from tab to tab add a class of `ui-state-persist` along with `ui-btn-active`. The markup for persistent tab bars is highlighted below.

Listing 3–9. *Persistent tab bars*

```
<!-- Movies tab bar -->
<div data-role="footer" class="tabbar" data-id="main-tabbar"
     data-position="fixed">
  <div data-role="navbar" class="tabbar">
    <ul>
      <li><a href="tabbar-movies.html"
            class="ui-btn-active ui-state-persist">Movies</a></li>
      <li><a href="tabbar-theatres.html">Theatres</a></li>
    </ul>
  </div>
</div>

<!-- Theatres tab bar -->
<div data-role="footer" class="tabbar" data-id="main-tabbar"
     data-position="fixed">
  <div data-role="navbar" class="tabbar">
    <ul>
      <li><a href="tabbar-movies.html">Movies</a></li>
      <li><a href="tabbar-theatres.html"
            class="ui-btn-active ui-state-persist">Theatres</a></li>
    </ul>
  </div>
</div>
```

Tab Bar with Custom Icons

Looking to add custom icons to your tab bars or toolbars? jQuery Mobile supports the addition of custom icons with minimal markup necessary. For example, in the tab bar example below (see Figure 3–13) we included several third-party icons from Glyphish.[1]

Figure 3–13. *Tab bar with custom Icons*

To support the addition of custom icons we need the addition of the `data-icon="custom"` attribute, some custom styling for positioning, and the id reference to associate each button with its style. These additions are highlighted in Listing 3–10 below.

---

[1] See `http://glyphish.com/`. Icons by Joseph Wain and licensed under the Creative Commons Attribution 3.0 United States License.

> **TIP:** This custom icon solution works equally well with toolbars too. In fact, by simply removing the text from the buttons creates a slim toolbar with custom icons!

Listing 3–10. *Tab bar with custom icons (ch3/tabbar-icons-custom.html)*

```
<!-- tab bar with custom icons -->
<div data-role="footer" class="ui-navbar-custom" data-position="fixed">
  <div data-role="navbar" class="ui-navbar-custom">
   <ul>
    <li><a href="#" id="home" data-icon="custom">Home</a></li>
    <li><a href="#" id="movies" data-icon="custom"
          class="ui-btn-active">Movies</a></li>
    <li><a href="#" id="theatres" data-icon="custom">Theatres</a></li>
   </ul>
  </div>
</div>


<style>
   .ui-navbar-custom .ui-btn .ui-btn-inner {
      font-size: 11px!important;
      padding-top: 24px!important;
      padding-bottom: 0px!important;
   }
   .ui-navbar-custom .ui-btn .ui-icon {
      width: 30px!important;
      height: 20px!important;
      margin-left: -15px!important;
      box-shadow: none!important;
      -moz-box-shadow: none!important;
      -webkit-box-shadow: none!important;
      -webkit-border-radius: none !important;
      border-radius: none !important;
   }
   #home .ui-icon {
      background: url(../images/53-house-w.png) 50% 50% no-repeat;
      background-size: 22px 20px;
   }
   #movies .ui-icon {
      background: url(../images/107-widescreen-w.png) 50% 50% no-repeat;
      background-size: 25px 17px;
   }
   #theatres .ui-icon {
      background: url(../images/15-tags-w.png) 50% 50% no-repeat;
      background-size: 20px 20px;
   }
</style>
```

Tab Bar with a Segmented Control

At this point we have seen examples of tab bars and segmented controls. How about merging the two together! We can utilize the persistent tab bar to help navigate our site and we can leverage the segmented control to display different views of our data. In the example below (see Figure 3–14) we have created a UI that allows the user to navigate between a home, movies, and theatres tab. When the user selects the movies tab we display the segmented control within the header to allow the user to help filter their movie listings. In this example we have completely removed the header text because the active tab highlights the title of our page. For the complete source code listing of this example, refer to ch3/tabbar-and-segmented-control.html.

Figure 3–14. *Tab bar with a segmented control*

Summary

In this chapter, we uncovered nearly every header and footer combination possible within jQuery Mobile. jQuery Mobile has an abundant set of components that greatly simplify navigation and data management requirements. We saw tab bar solutions that provide the ability to switch between different views within the app. We reviewed several toolbar configurations that help manage objects on the current screen. And we added segmented controls to give users access to different perspectives of application data. Additionally, each component is flexible in regards to their appearance. Each component is themable and we can style our buttons with icons, text, or a combination of both. In the next chapter, we will review all possible button styling options and also look at the components we may use for form-based development within jQuery Mobile.

Chapter 4

Form Elements and Buttons

Mobile applications must support an efficient user experience. For this reason, it is rare to see mobile apps with numerous form fields. In fact, the less interaction our apps require of our users, the more efficient both the users and the apps will become. The mobile Web is slowly adopting device APIs[1] that allow developers to collect an abundance of information with minimal user interaction. For instance, 74% of mobile developers are using Geolocation within their apps[2]. Geolocation allows us to gather the user's country, state, city, zip code, and address information with the simple tap of an acknowledgement button. Although these device APIs are making the user experience more efficient, for users whose devices don't support geolocation, we still need to capture data the traditional way with form fields.

In this chapter, we will start with the most popular mobile UI component, the button. Buttons can be styled and configured in many ways. We will see examples of buttons styled with text, icons, and a combination of both.

Next, we will take a detailed look at every standard HTML form component and identify common use cases they solve really well. You will be pleasantly surprised that each form component is automatically optimized by jQuery Mobile—a feature that conveniently provides a unified user experience across all devices. We will also review the jQuery Mobile data attributes that are unique to each form element, and see code examples in which we modify these attributes to configure and style our forms. Furthermore, we will review the plugins that are associated with each form component and see how we can leverage the plugin API to dynamically create, enhance, and update our own components when users require a more dynamic experience.

---

[1] See http://www.w3.org/2009/dap/.

[2] See http://www.webdirections.org/sotmw2011/.

Lastly, we will explore the features of the Mobiscroll plugin, which provides an elegant and flexible interface for date pickers, search filters, or custom lists.

Buttons

Buttons are the most commonly used control within mobile apps because they provide a very efficient user experience. We have already seen buttons used in many examples, including our dialogs, action sheets, segmented controls, and header. jQuery Mobile buttons come in many flavors. We have link buttons, form buttons, image buttons, icon-only buttons, and buttons combined with text and icons. As expected, jQuery Mobile buttons are all styled consistently. Whether you have a link button or a form-based button, the framework will style them identically. As we review these buttons we will also identify common use cases for each type of button.

Link Buttons

Link buttons are the most commonly used type of button. Whenever you need to style an ordinary link as a button, add the data-role="button" attribute to the link (see Figure 4–1).

Figure 4–1. *Link buttons*

By default, buttons within the content section of a page are styled as block-level elements so they will fill the entire width of their outer container. However, if you want a more compact button that is only as wide as the text and icons inside, add the data-inline="true" attribute (see Listing 4–1).

Listing 4–1. *Link buttons (ch4/link-buttons.html)*

```
<a href="#" data-role="button">Link button</a>
<a href="#" data-role="button" data-inline="true">Disagree</a>
<a href="#" data-role="button" data-inline="true">Agree</a>
```

> **NOTE:** If you want buttons to sit side-by-side and consume the entire width of the screen, use a 2-column grid. We will explore flexible grid layouts in more detail in Chapter 6. Specifically, for a 2-column grid layout refer to Listing 6-2.

Form Buttons

Form-based buttons (see Listing 4–2) are actually easier to style than link-based buttons because no modifications are required on your part. For simplicity, the framework automatically converts any `button` or `input` element into a mobile-styled button for you (see Figure 4–2).

Listing 4–2. *Form buttons (ch4/form-buttons.html)*

```
<button type="submit">Button element</button>
<input type="button" value="button" />
<input type="submit" value="submit" />
<input type="reset" value="reset" />
```

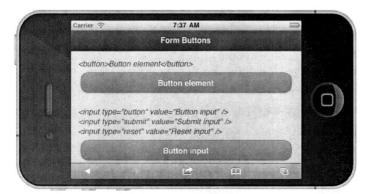

Figure 4–2. *Form buttons*

> **TIP** If you want to disable the automatic initialization of form buttons or any other control, you may add the `data-role="none"` attribute to the element and jQuery Mobile will not enhance the control:
>
> `<button data-role="none">Button element</button>`

Image Buttons

Styling images as buttons requires minimal effort on your part. When wrapping an image with an anchor tag, no modifications are necessary (see Figure 4–3 and its related code in Listing 4–3). However, when attaching an image to an input element you will need to add the `data-role="none"` attribute.

Figure 4–3. *Image buttons*

Listing 4–3. *Image buttons (ch4/image-buttons.html)*

```
<!-- Image buttons -->
<input type="image" src="cloud.png" data-role="none" />
<a href="#"><img src="cloud.png"></a>
```

Styling Buttons with Icons

jQuery Mobile includes a set of standard icons that are commonly used in mobile applications, which includes a single white icon sprite that has a semi-transparent black circle behind the icon to ensure a good contrast on any background color (see Figure 4–4).

Figure 4–4. *Buttons with standard icons*

An icon can be added to any button by adding the data-icon attribute and specifying which icon to display (see Listing 4–4).

Listing 4–4. *Buttons with icons (ch4/icon-buttons-standard.html)*

```
<!-- Buttons with standard icons.  Refer to Table 4-1 for icon list. -->
<input type="button" value="Delete" data-icon="delete"/>
<a href="#" data-role="button" data-icon="plus">Button link</a>
<button data-icon="minus">Button element</button>
```

Table 4–1 contains each data-icon attribute value and its corresponding icon image. Each attribute value has an associated image except for data-icon="custom". We will see an example of integrating with custom icons in the next section.

Table 4–1. *data-icon listing*

data-icon	Image
plus	✚
minus	▬
delete	✖
arrow-r	❯
arrow-l	❮
arrow-u	︿
arrow-d	﹀
check	✔
gear	✿
refresh	↻
forward	↻
back	↺
grid	⠿
star	★

data-icon	Image
alert	⚠
info	ⓘ
home	⌂
search	🔍
custom	

Icon-only Buttons

Icon-only buttons are commonly used within headers, toolbars, and tab bars because they consume very little real estate (see Figure 4–5).

Figure 4–5. *Icon-only buttons*

In the last chapter we saw several examples of icon-only buttons. We initially saw a "plus" icon in Figure 3-4 that allowed users to tap the "add" icon to create a new movie review. We also saw icon-only buttons used within our toolbar (see Figure 3-10) and tab bars (see Figure 3-12) to help express the meaning of each button. To create an icon-only button add the `data-iconpos="notext"` attribute to the button (see Listing 4–5).

Listing 4–5. *Icon-only buttons (ch4/icon-only-buttons.html)*

```
<a href="" data-role="button" data-icon="plus" data-iconpos="notext"></a>
<button data-icon="search" data-iconpos="notext">Search</button>
```

NOTE: The semi-transparent black circle behind each white icon ensures a good contrast on any background color and works well with the jQuery Mobile theming system. For instance, in the image below, the icons in the first row are styled with data-theme="a" and the icons in the second row have the data-theme="c" styling. To maintain visual consistency, it is recommended to create a white icon 18 × 18 pixels saved as a PNG-8 with alpha transparency.

Icon Positioning

By default, icons will be left-aligned (see Figure 4–6). However, you may explicitly align icons to any side by adding the data-iconpos attribute to the button with its value corresponding to the side of alignment (see Listing 4–6).

Figure 4–6. *Icon positioning*

Listing 4–6. *Icon-only buttons (ch4/icon-positioning.html)*

```
<a href="#" data-role="button" data-icon="arrow-u" data-iconpos="top">
<a href="#" data-role="button" data-icon="arrow-l" data-iconpos="left">
<a href="#" data-role="button" data-icon="arrow-r" data-iconpos="right">
<a href="#" data-role="button" data-icon="arrow-d" data-iconpos="bottom">
```

Buttons with Custom Icons

Remember when we added custom Glyphish icons to our tab bar back in Figure 3-13? We can integrate buttons with custom icons in the same manner (see Figure 4–7).

Figure 4–7. *Custom icons*

However, with buttons we can apply a more simplified solution as shown in Listing 4–7. Two steps are necessary to add custom icons to your buttons:

1. Add a `data-icon` attribute to the link. The value of this attribute must uniquely identify the custom icon. For example, `data-icon="my-custom-icon"`.

2. Create a CSS class attribute that sets the background source for our custom image. The name of the class attribute must be named ".ui-icon-<data-icon-value>". For example, if our data-icon value was "my-custom-icon", our new CSS class attribute would be ".ui-icon-my-custom-icon".

Listing 4–7. *Custom icon integration (ch4/icon-buttons-custom.html)*

```
<style>
  .ui-icon-custom1 {
      background:url(data:image/png;base64,iVBORw0...)50% 50% no-repeat;
      background-size: 14px 14px;
  }
</style>
<a href="#" data-role="button" data-icon="custom1">Custom</a>
```

> **TIP:** The background source for our custom image was loaded with the data URI scheme. This can be a performant alternative to loading small images externally. For instance, by including the custom image in-line we have eliminated an HTTP request. However, the main disadvantage of this technique is the size of the base64 encoded string is 1/3 times larger than the original image. To see the complete base64 encoded string refer to the source code listing in ch4/icon-buttons-custom.html.

Grouping Buttons

Thus far, every button example shown had each button segregated from the others. However, if you want to group your buttons together, you can wrap your buttons within a control group. For example, our segmented control examples in Chapter 3 were grouped this way (see Figure 4–8).

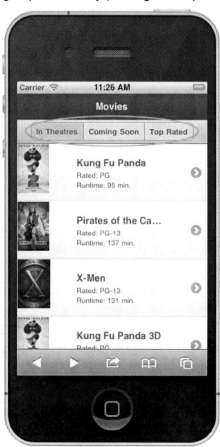

Figure 4–8. *Grouping Buttons*

To get this effect, wrap a group of buttons in a container with the `data-role="controlgroup"` attribute (see Listing 4–8).

Listing 4–8. *Grouping buttons (ch3/header-segmented-control.html)*

```
<div data-role="controlgroup" data-type="horizontal">
    <a href="#" data-role="button">In Theatres</a>
    <a href="#" data-role="button">Coming Soon</a>
    <a href="#" data-role="button">Top Rated</a>
</div>
```

By default, the framework will group the buttons vertically, remove all margins, and add borders between the buttons. Additionally, to visually enhance the group, the first and last elements will be styled with rounded corners.

Because buttons are positioned vertically by default, we can style them horizontally with the addition of the `data-type="horizontal"` attribute. Unlike vertical buttons that consume the entire width of their outer container, horizontal buttons are only as wide as their content.

> **CAUTION:** When grouping buttons horizontally, the control group will wrap when its width extends beyond the width of the screen.

Theming Buttons

Buttons, like all jQuery Mobile components, will inherit the theme from their parent container. Furthermore, when you need to style buttons with different colors you can apply the theme of your choice to any button with the addition of the `data-theme` attribute (see Listing 4–9).

Listing 4–9. *Theming buttons (ch2/action-sheet2.html)*

```
<a href="#home" data-role="button" data-theme="b">YouTube</a>
<a href="#home" data-role="button" data-theme="b">Facebook</a>
<a href="#home" data-role="button" data-theme="b">Email</a>
<a href="#home" data-role="button" data-theme="c">Cancel</a>
```

For instance, in our dialog and action sheet examples we styled our buttons according to the "Dialog UX Guidelines" in Chapter 2 for improved usability (see Figure 4–9).

Figure 4–9. *Theming Buttons*

Dynamic Buttons

The button plugin is the widget that automatically enhances native buttons. We can leverage this plugin to dynamically create, enable, and disable buttons. If you need to create buttons dynamically in code there are two options available. You can create buttons dynamically with a markup-driven approach or by explicitly setting the options on the button plugin.

In the markup-driven solution, we create the jQuery Mobile markup for the new button, append it to the content container, and enhance it (see Listing 4–10).

Listing 4–10. *Create dynamic button with markup-driven options (ch4/dynamic-buttons.html)*

```
// Add link button to content container and enhance it
$( '<a href="#" data-role="button" data-icon="star" id="b1">Star</a>' )
        .appendTo( ".ui-content" )
        .button();

// Add form button after the first button and enhance it
```

```
$( '<input type="submit" id="b2" value="Button 2" data-theme="a" />' )
        .insertAfter( "#b1" )
        .button();
```

For the option-driven solution, we create a native link, insert the button onto the page, and then apply our button enhancements (see Listing 4–11).

Listing 4–11. *Create dynamic button with plugin-driven options (ch4/dynamic-buttons.html)*

```
// Create a new button, insert it after button 2, and enhance it.
$( '<a href="#">Home</a>' )
        .insertAfter( "#b2" )
        .button({
                'icon':'home',
                'inline': true,
                'shadow': true,
                'theme': 'b'
        });
```

In our last example, we create multiple form buttons and instead of calling the button plugin individually for each button we enhance them all with a single call by triggering the "create" method on the page container (see Listing 4–12). The button plugin also exposes enable and disable methods that we can leverage to dynamically enable and disable buttons as shown in Listing 4–12.

Listing 4–12. *Create buttons and dynamically disable/enable them (ch4/dynamic-buttons.html)*

```
// Create multiple form buttons
$( '<button id="button3">Button3</button>' ).insertAfter( "#button2" );
$( '<button id="button4">Button4</button>' ).insertAfter( "#button3" );

// Enhance all widgets on the page
$.mobile.pageContainer.trigger( "create" );

// Disable form button
$( "#button3" ).button( "disable" );

// Enable form button
$( "#button3" ).button( "enable" );
```

> **TIP:** Triggering the "create" method on the page container will enhance all components on the page: $.mobile.pageContainer.trigger("create"); This is a convenient method when you need to enhance multiple page components at once.

Button Options

The button plugin, which is used by the framework to dynamically enhance buttons, has the following options:

corners *boolean*
        default: true

By default, buttons will have rounded corners. Setting this option to false will remove the rounded corners. This option is also exposed as a data attribute: `data-corners="false"`.

```
$( "#button1" ).button({ corners: false });
```

icon *string*
    default: null

Sets the icon for the button. This option is also exposed as a data attribute: `data-icon="plus"`.

```
$( "#button1" ).button({ icon: "home" });
```

iconpos *string*
    default: "left"

Sets the icon position. The possible values are: "left", "right", "top", "bottom", and "notext". The "notext" value will display the button as an icon-only button with no text. This option is also exposed as a data attribute: `data-iconpos="notext"`.

```
$( "#button1" ).button({ iconpos: "notext" });
```

iconshadow *boolean*
    default: true

When true, the framework will add a drop shadow to the icon. This option is also exposed as a data attribute: `data-iconshadow="false"`.

```
$( "#button1" ).button({ iconshadow: false });
```

initSelector *CSS selector string*
    default: "button, [type='button'], [type='submit'], [type='reset'], [type='image']"

The initSelector is used to define the selectors (element types, data roles, etc.) that are used to trigger the automatic initialization of the widget plugin. For instance, all elements that are matched by the default selector will be enhanced by the button plugin. To override this selector, bind to the mobileinit event and update the selector as necessary:

```
$( document ).bind( "mobileinit", function(){
$.mobile.button.prototype.options.initSelector = "...";
});
```

inline *boolean*
    default: false

If set to true, this will make the button appear as an inline button. By default, buttons will consume the entire width of their container. In contrast, Inline buttons only consume the width of their text. This option is also exposed as a data attribute: `data-inline="true"`.

```
$( "#button1" ).button({ inline: true });
```

shadow *boolean*
> default: true

> By default, buttons will have a drop shadow applied. Setting this option to false will remove the drop shadow. This option is also exposed as a data attribute: data-shadow="false".

```
$( "#button1" ).button({ shadow: false });
```

Button Methods

The button plugin has the following methods:

enable: enable a disabled button
```
$( "#button1" ).button( "enable" );
```

disable: disable a button
```
$( "#button1" ).button( "disable" );
```

Button Events

The button plugin supports the following events:

create triggered when a button is created

> This event is triggered when a custom button is created. It is not used to create a custom button.

```
$( '<a href="#" id="button2">Button2</a>' )
      .insertAfter( "#button1" )
      .button({
            theme: 'a',
            create: function(event) {
                  console.log( "Creating button..." );
            }
      })
```

Form Elements

jQuery Mobile will enhance all native form elements to make them more attractive and usable on mobile devices. However, older browsers that do not support these enhancements will progressively fall back to native elements to maintain a usable experience.

Form Basics

Methods for building form-based applications within jQuery Mobile are very similar to those we have traditionally used to build forms on the Web. Although an `action` and method attribute should be specified for clarity, they are not required. By default, the action will default to the current page's relative path, which can be found with `$.mobile.path.get()` and an unspecified `method` will default to "get".

When forms are submitted, they will transition to their subsequent page with the default "slide" transition. However, we may configure our form transition behavior with the same data attributes we used previously to manage our links (see Listing 4–13).

Listing 4–13. *Submitting forms (ch4/form-request.html)*

```
<form action="/save.html" method="post" data-transition="pop">
        <label for="email">Email:</label>
        <input type="email" name="email" id="email" value="" />
        <button type="submit" name="submit">Submit</button>
</form>
```

We can add the following attributes to our form element to manage transitions or to disable Ajax:

- `data-transition="pop"`
- `data-direction="reverse"`
- `data-ajax="false"`

> **CAUTION:** It is important to ensure that `id` attributes for each form are unique across your entire site. As mentioned previously, when transitioning jQuery Mobile will load the "from" and "to" pages into the DOM at the same time to complete a smooth transition. To avoid any collisions, form id's must be unique.

TIP: When building forms it is recommended to semantically associate each form field with its corresponding label. The label's **for** attribute and the input's **id** attribute establish this relationship:

```
<label for="name">Name:</label>

<input type="text" name="name" id="name" value="" />
```

This association creates 508-compliant applications that are accessible to assistive technologies. Accessibility is often required by government or state agencies. You can test your mobile application for compliance with the WAVE[3] tool.

Text Inputs

Text inputs are the most cumbersome form field to work with on mobile devices. Unless you are a world texting champion, entering text on a physical or virtual QWERTY keyboard is inefficient. This is why it is valuable to automatically collect as much user information as possible. As mentioned earlier, device APIs can help simplify this user experience. Although it is a good goal to minimize these tedious tasks, there are times when we must collect user feedback with text inputs. The most common text form fields are shown in Figure 4–10.

---

[3] See http://wave.webaim.org/.

Figure 4–10. *Text inputs*

From a developer perspective, we can create jQuery Mobile forms and text inputs with no additional markup necessary (see Listing 4–14). Optionally, we can choose an appropriate theme for our text inputs by adding the data-theme attribute to our input element to enhance form field contrast.

Listing 4–14. *Text inputs (ch4/text-inputs.html)*

```
<input type="text" name="text" value="" id="text" placeholder="Text"/>
<input type="number" name="number" value="" id="number" />
<input type="email" name="email" value="" id="email" data-theme="d" />
<input type="url" name="url" value="" id="url" />
<input type="tel" name="tel" value="" id="tel" />
<input type="search" name="search" value="" id="search" />
<textarea cols="40" rows="8" name="textarea" id="textarea"></textarea>
```

> **TIP:** To hide labels in an accessible way attach the ui-hidden-accessible style to the element. For instance, we applied this technique to the search field in Figure 4-10. This will gracefully hide the label while preserving 508 compliance:
>
> ```
> <label for="search" class="ui-hidden-accessible">Search</label>
>         <input type="search" id="search" placeholder="Search" />
> ```

When building forms, it is important to associate the input field with its semantic type. This association has two advantages. First, when the input field receives focus it prompts the user with the appropriate keyboard. For instance, a field that is specified as type="number" will automatically prompt a numeric keyboard (see Figure 4–11). Likewise, a field that is mapped with type="tel" will prompt a telephone-specific keyboard (see Figure 4–12).

Figure 4–11. *Numeric keyboard*

Figure 4–12. *Telephone keyboard*

Additionally, this specification allows the browser to apply validation rules that are applicable for the field type. The browser support for automatic validation when submitting forms is still minimal but will improve over time. For a complete listing of mobile input types and attributes refer to Peter-Paul Koch's, "Input tests for mobile"[4]. It shows all available mobile input types and attributes with their associated browser support.

Another feature that is well supported across most mobile browsers is the `placeholder` attribute. This attribute adds a hint or label to the text input and automatically disappears when the field receives focus (see Listing 4–14).

> **NOTE:** The search field (`type="search"`) is styled and behaves slightly different than the other input types. It contains a left-aligned "search" icon, its corners are pill-shaped, and when users enter text a "delete" icon will appear right-aligned to help clear the field.

Dynamic Text Inputs

The `textinput` plugin is the widget that automatically enhances text inputs and text areas. We can leverage this plugin to dynamically create, enable, and disable text inputs (see Listing 4–15).

Listing 4–15. *textinput plugin examples (ch4/dynmic-text-input.html)*

```
// Create text input with markup-driven options
$( '<input type="text" name="text1" value="" data-theme="c" />' )
        .insertAfter( "#firstName" )
        .textinput();

// Create text input with plugin-driven options
$( '<input type="text" name="text2" id="text2" value="" />' )
        .insertAfter( "#text1" )
        .textinput({
                theme: 'c'
        });

// Disable text input
$( "#text1" ).textinput( "disable" );

// Enable text input
$( "#text1" ).textinput( "enable" );
```

---

[4] See http://www.quirksmode.org/html5/inputs_mobile.html.

Text Input Options

The textinput plugin has the following options:

initSelector *CSS selector string*
>   **default:** "input[type='text'], input[type='search'], :jqmData(type='search'),
>   input[type='number'], :jqmData(type='number'), input[type='password'],
>   input[type='email'], input[type='url'], input[type='tel'], textarea"

>   The initSelector is used to define the selectors (element types, data roles, etc.)
>   that are used to trigger the automatic initialization of the widget plugin. For
>   instance, all elements that are matched by the default selector will be enhanced
>   by the textinput plugin. To override this selector, bind to the mobileinit event
>   and update the selector as necessary:

>    ```
>    $( document ).bind( "mobileinit", function(){
>            $.mobile.textinput.prototype.options.initSelector = "...";
>    });
>    ```

theme *string*
>   **default:** null. Inherited from parent.

>   Sets the theme swatch color scheme for the text element. This is a letter from a
>   to z that maps to the swatches included in your theme. By default, all elements
>   will inherit the same swatch color as their parent container if not explicitly set.
>   This option is also exposed as a data attribute: data-theme="a".

>    ```
>    $( "#text1" ).textinput({ theme: "a" });
>    ```

Text Input Methods

The textinput plugin has the following methods:

enable: *enable a disabled textinput or textarea.*

>    ```
>    $( "textarea" ).textinput( "enable" );
>    ```

disable: *disable a textinput or textarea.*

>    ```
>    $( "textarea" ).textinput( "disable" );
>    ```

Text Input Events

The textinput plugin supports the following events:

create *triggered when a text input is created*

This event is triggered when a custom text input is created. It is not used to create a custom input.

```
$( '<input type="text" name="text2" id="text2" value="" />' )
        .textinput({
                theme: 'c',
                create: function(event) {
                        console.log( "Creating text input..." );
                }
        })
        .insertAfter( "#text1" );
```

Select Menus

The jQuery Mobile framework will automatically enhance all native select elements with no additional markup required (see Listing 4–16).

Listing 4–16. *Native select menu (ch4/select-menu-native.html)*

```
<label for="genre">Genre:</label>
<select name="genre" id="genre">
        <option value="action">Action</option>
        <option value="comedy">Comedy</option>
        <option value="drama">Drama</option>
</select>
```

This transformation will replace the original select with a jQuery Mobile styled button that contains a down-arrow icon that is right-aligned. By default, tapping this select button will launch the native select picker for the OS (see Figure 4–13). Alternatively, as we will see in the next section, we can configure jQuery Mobile to display custom select menus.

Figure 4–13. *Select menu*

After users make their selection, the select button will display the value of the chosen option(s). If the text value is too large for the button, the text will be truncated and a trailing ellipsis will be shown. Additionally, multi-select buttons will display a count bubble or badge after selecting more than one option (see Figure 4–13). This is a visual effect that highlights the number of selected options.

> **CAUTION:** Some mobile platforms do not support the multi-select feature when creating a select menu with the `multiple="multiple"` attribute. Therefore, using custom menus when this behavior is necessary is recommended.

Custom Select Menus

As an alternative to the natively rendered options list, we may opt to have our select menus rendered in a custom HTML/CSS view (see Figure 4–14).

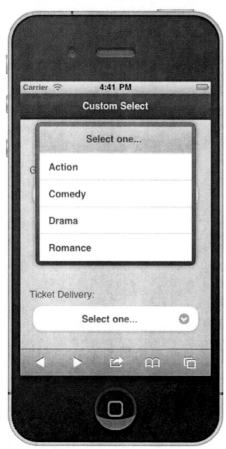

Figure 4–14. *Custom select menu*

For this view, add a `data-native-menu="false"` attribute to the select element (see Listing 4–17).

Listing 4–17. *Custom Select menu (ch4/select-menu-cstom.html)*

```
<label for="genre">Genre:</label>
<select name="genre" id="genre" data-native-menu="false" data-theme="a">
        <option value="">Select one...</option>
        <option value="action">Action</option>
        <option value="comedy">Comedy</option>
        <option value="drama">Drama</option>
</select>
```

A breakdown of custom versus native advantages is listed below.

Custom advantages:

- Provides a unified user experience across all devices.

- The custom menu provides universal support for the multi-select option list.

- Adds an elegant way to handle placeholder options. We will review placeholder options in the next section.

- The custom menus are themable (see Listing 4–17).

Custom disadvantages:

- Not as performant as the native-rendered select menu. This will be more apparent when comparing menus containing many options.

> **NOTE:** You may also leverage the `Mobiscroll`[5] plugin as another alternative for customized select menus. At the end of the chapter we include several examples of this plugin as we demonstrate its usage for a date picker and search filter.

Placeholder Options

A placeholder is a feature that is unique to custom select menus. A placeholder provides three benefits:

1. A placeholder requires users to make a selection. By default, the first option in the list will be selected if no placeholder has been configured.

2. A placeholder can be used to display hint text for the unselected select button (see Figure 4–14). For instance, the unselected Ticket Delivery field is shown with the placeholder text of "Select one...".

3. A placeholder also appears as the header when the options list is displayed (see Figure 4–14).

---

[5] See `http://code.google.com/p/mobiscroll/`.

We can configure a placeholder in three ways:

1. We can add text to an option without a value.
    ```
    <option value="">Select one...</option>
    ```

2. We can add the `data-placeholder="true"` attribute to an option when it contains text and a value:
    ```
    <option value="null" data-placeholder="true">Select one...</option>
    ```

3. When you want to make the field required without hint text or a header, use an empty option:
    ```
    <option value=""></option>
    ```

Dynamic Select Menus

The `selectmenu` plugin is the widget that automatically enhances a select menu. With this plugin we can dynamically create, enable, disable, open, and close select menus (see Listing 4–18).

Listing 4–18. *Dynamic select menus (ch4/dynamic-select-menu.html)*

```
// Create select menu with markup-driven options
$( '<select name="select1" id="select1" data-theme="e">...</select>' )
       .insertAfter( "#foo" )
       .selectmenu();

// Create select menu with plugin-driven options
$( '<select name="select2" id="select2">...</select>' )
       .insertAfter( "#select1" )
       .selectmenu({
              theme: "e",
              overlayTheme: "c",
              disabled: false,
              nativeMenu: false
       });
```

Select Menu Options

The `selectmenu` plugin has the following options:

corners *boolean*
> **default:** true
>
> Like other button types, select menu buttons will have rounded corners by default. Setting this option to false will remove the rounded corners. This option is also exposed as a data attribute: `data-corners="false"`.
>
> ```
> $( "#select1" ).selectmenu({ corners: false });
> ```

disabled *boolean*
　　default: false

　　Disables the element. The selectmenu plugin also has enable and disable
　　methods to dynamically enable and disable the control.

```
$( "#select1" ).selectmenu({ disabled: true });
```

hidePlaceholderMenuItems *boolean*
　　default: true

　　By default, placeholder menu items will be hidden from view when the select
　　menu is open. To allow the placeholder item to be selectable, set this value to
　　false.

```
$( "#select1" ).selectmenu({ hidePlaceholderMenuItems: false });
```

icon *string*
　　default: "arrow-d"

　　Sets the icon for the select button. This option is also exposed as a data
　　attribute: data-icon="plus".

```
$( "#select1" ).selectmenu({ icon: "plus" });
```

iconpos *string*
　　default: "right"

　　Sets the icon position. The possible values are: "left", "right", "none", and
　　"notext". The "notext" value will display the select as an icon-only button with
　　no placeholder text. The "none" value will remove the icon completely. This
　　option is also exposed as a data attribute: data-iconpos="none".

```
$( "#select1" ).selectmenu({ iconpos: "notext" });
```

iconshadow *boolean*
　　default: true

　　When true, the framework will add a drop shadow to the icon. This option is also
　　exposed as a data attribute: data-iconshadow="false".

```
$( "#select1" ).selectmenu({ iconshadow: false });
```

initSelector *CSS selector string*
　　default: "select:not(:jqmData(role='slider'))"

　　The initSelector is used to define the selectors (element types, data roles, etc.)
　　that trigger the automatic initialization of the widget plugin. For instance, all
　　elements that are matched by the default selector will be enhanced by the
　　selectmenu plugin. To override this selector, bind to the mobileinit event and
　　update the selector as necessary:

```
$( document ).bind( "mobileinit", function(){
        $.mobile.selectmenu.prototype.options.initSelector = "..";
});
```

`inline` *boolean*
      default: false

If set to true, this will make the select button appear as an inline button. By default, select buttons will consume the entire width of their container. In contrast, inline buttons only consume the width of their placeholder text. This option is also exposed as a data attribute: `data-inline="true"`.

`$( "#select1" ).selectmenu({ inline: true });`

`nativeMenu` *boolean*
      default: true

By default, select buttons will launch the native select picker for the OS. To render the select menu in a custom HTML/CSS view, set this value to false. This option is also exposed as a data attribute: `data-native-menu="false"`.

`$( "#select1" ).selectmenu({ nativeMenu: false });`

`shadow` *boolean*
      default: true

By default, select buttons will have a drop shadow applied. Setting this option to false will remove the drop shadow. This option is also exposed as a data attribute: `data-shadow="false"`.

`$( "#select1" ).selectmenu({ shadow: false });`

`theme` *string*
      default: null. Inherited from parent.

Sets the theme swatch color scheme for element. This is a letter from a to z that maps to the swatches included in your theme. By default, this will inherit the same swatch color as its parent container. This option is also exposed as a data attribute: `data-theme="a"`.

`$( "#select1" ).selectmenu({ theme: "a" });`

Select Menu Methods

The `selectmenu` plugin has the following methods:

`enable:` *enable a disabled select menu*

      `$( "#select1" ).selectmenu( "enable" );`

`disable:` *disable a select menu.*

      `$( "#select1" ).selectmenu( "disable" );`

open: *open a closed select menu. This method only works on a custom select.*

```
$( "#select1" ).selectmenu( "open" );
```

close: *close an open select menu. This method only works on a custom select.*

```
$( "#select1" ).selectmenu( "close" );
```

refresh: *update the custom select menu.*

This updates the custom select menu to reflect the native select element's value. For instance, if the selectedIndex of the native select is updated we can call "refresh" to rebuild the custom select. If you pass a true argument, you can force a refresh and a rebuild to occur.

```
// Select the third menu option and refresh the menu
        var myselect = $( "#select1" );
        myselect[0].selectedIndex = 2;
        myselect.selectmenu( "refresh" );

        // refresh and rebuild the list
        myselect.selectmenu( "refresh", true );
```

Select Menu Events

The selectmenu plugin supports the following events:

create: triggered when a select menu is created

This event is triggered when a custom select menu is created. It is not used to create a custom element.

```
$( '<select name="select2" id="select2">...</select>' )
        .insertAfter( "#select1" )
        .selectmenu({
                create: function(event) {
                        console.log( "Creating select menu..." );
                }
        });
```

Radio Buttons

Radio buttons limit the user's selection to a single item (see Figure 4–15). For instance, choosing an application setting from among several options is commonly accomplished with radio buttons, which are preferred for their simplicity and ease of use. Users may tap anywhere on the radio button to make their selection, and jQuery Mobile will automatically update the underlying form control.

Figure 4–15. *Radio buttons*

In Listing 4–19 we added three additional attributes to help style and position our radio buttons. The first attribute, `data-role="controlgroup"`, elegantly groups the buttons together with rounded corners. The second attribute, `data-type="horizontal"`, overrides the default positioning—vertical—and displays the buttons horizontally. Lastly, we themed our buttons. By default, radio buttons will inherit the theme of their parent control. However, if you want to apply an alternate theme to your radio buttons you may add the `data-theme` attribute to the label of the corresponding radio button.

Listing 4–19. *Horizontal radio buttons (ch4/radio-buttons.html)*

```
<fieldset data-role="controlgroup" data-type="horizontal">
    <legend>Map view:</legend>
    <input type="radio" name="map" id="map1" value="Map" />
    <label for="map1" data-theme="b">Map</label>

    <input type="radio" name="map" id="map2" value="Satellite" />
    <label for="map2" data-theme="b">Satellite</label>

    <input type="radio" name="map" id="map3" value="Hybrid"  />
```

```
            <label for="map3" data-theme="b">Hybrid</label>
</fieldset>
```

> **Caution:** Horizontal radio buttons will wrap if their container is not wide enough to display them
> on a single row. You may reduce their font size if wrapping is an issue: .ui-controlgroup-
> horizontal .ui-radio label {font-size: 13px !important;}

Dynamic Radio Buttons

The checkboxradio plugin is a reusable widget that automatically enhances both radio
buttons and checkboxes. With this plugin we can dynamically create, enable, disable,
and refresh our radio buttons (see Listing 4–20).

Listing 4–20. *Dynamic radio buttons (ch4/dynamic-radio-buttons.html)*

```
// Create radio buttons with markup-driven options
$( '<fieldset data-role="controlgroup">
        <legend>Map view:</legend>
        <input type="radio" name="map" id="map1" value="Map" />
        <label for="map1" data-theme="c">Map</label>...</fieldset>')
        .insertAfter( "#radio1" );
$.mobile.pageContainer.trigger( "create" );

// Create radio buttons with plugin-driven options
$( '<fieldset data-role="controlgroup">
        <legend>Map view:</legend>
        <input type="radio" name="map" id="map1" value="Map" />
        <label for="map1">Map</label>...</fieldset>' )
        .insertAfter( "#radio1" );
$( "#map1" ).checkboxradio({ theme: "e" });
$( "#map2" ).checkboxradio({ theme: "e" });
$.mobile.pageContainer.trigger( "create" );
```

Checkbox and Radio Button Options

The checkboxradio plugin has the following options:

initSelector *CSS selector string*
        default: "input[type='checkbox'],input[type='radio']"

        The initSelector is used to define the selectors (element types, data roles, etc.)
        that are used to trigger the automatic initialization of the widget plugin. For
        instance, all elements that are matched by the default selector will be enhanced
        by the checkboxradio plugin. To override this selector, bind to the mobileinit
        event and update the selector as necessary:

```
$( document ).bind( "mobileinit", function(){
    $.mobile.checkboxradio.prototype.options.initSelector = "...";
});
```

theme *string*

> **default:** null. Inherited from parent.
>
> Sets the theme swatch color scheme for the checkbox or radio button. This is a letter from a to z that maps to the swatches included in your theme. By default, this will inherit the same swatch color as its parent container. This option is also exposed as a data attribute: data-theme="a".
>
> ```
> $( "#element1" ).checkboxradio({ theme: "a" });
> ```

Checkbox and Radio Button Methods

The checkboxradio plugin has the following methods:

enable: *enable a disabled checkbox or radio button.*

> ```
> $("#element1" ).checkboxradio( "enable" );
> ```

disable: disable a checkbox or radio button.

> ```
> $( "#element1" ).checkboxradio( "disable" );
> ```

refresh: *update the custom checkbox or radio button*

> This updates the custom checkbox or radio button to reflect the native element's value. For instance, we can dynamically check a radio button and call "refresh" to rebuild the enhanced control.
>
> ```
> // Dynamically set a checkbox or radio element and refresh it.
> $( "#elem1" ).attr( "checked", true ).checkboxradio( "refresh" );
> ```

Checkbox and Radio Button Events

The checkboxradio plugin supports the following events:

create: triggered when a checkbox or radio button is created

> This event is triggered when a custom checkbox or radio button is created. It is not used to create a custom element.
>
> ```
> $( '#element1' )
>         .checkboxradio({
>                 theme: "e",
>                 create: function(event) {
>                         console.log( "Creating new element..." );
>                 }
> });
> ```

Checkboxes

Checkboxes are a common form control for allowing users to select multiple values from a listing of many choices (see Figure 4–16). Users may tap anywhere on the checkbox button to make a selection, and jQuery Mobile will automatically update the underlying form control.

Figure 4–16 *Checkboxes*

The markup for styling and positioning checkboxes is identical to what we previously used for radio buttons (see Listing 4–21). Again, we add three additional attributes to help style and position our checkboxes. The first attribute, `data-role="controlgroup"`, elegantly groups the checkbox elements together with rounded corners. The second attribute, `data-type="horizontal"`, overrides the default vertical positioning of the buttons and displays them horizontally. Lastly, we themed our buttons. By default, checkboxes will inherit the theme of their parent control. However, if you want to apply an alternate theme you may add the `data-theme` attribute to the label of the corresponding checkbox.

Listing 4–21. *Horizonal checkboxes (ch4/checkboxes.html)*

```
<fieldset data-role="controlgroup" data-type="horizontal">
        <legend>Genre:</legend>
        <input type="checkbox" name="genre" id="c1" />
        <label for="c1" data-theme="c">Action</label>

        <input type="checkbox" name="genre" id="c2" />
        <label for="c2" data-theme="c">Comedy</label>

        <input type="checkbox" name="genre" id="c3" />
        <label for="c3" data-theme="c">Drama</label>
</fieldset>
```

> **CAUTION:** Horizontal checkboxes will wrap if their container is not wide enough to display them
> on a single row. You may reduce their font size if wrapping is an issue:
>
> ```
>     .ui-controlgroup-horizontal .ui-checkbox label {
>         font-size: 11px !important;                                    }
> ```

Dynamic Checkboxes

Again, the checkboxradio plugin is the widget that automatically enhances both
checkboxes and radio buttons. With this plugin we can dynamically create, enable,
disable, and refresh our checkboxes (see Listing 4–22). The complete listing of the
checkboxradio plugin was previously documented in the "Dynamic Radio Buttons"
Section. The same API is reusable for both radio buttons and checkboxes.

Listing 4–22. *Dynamic checkboxes (ch4/dynamic-checkboxes.html)*

```
// Create checkboxes with markup-driven options
$( '<fieldset data-role="controlgroup">
        <legend>Genre:</legend>
        <input type="checkbox" name="genre" id="c1" />
        <label for="c1" data-theme="c">Action</label>...</fieldset>' )
        .insertAfter( "#element1" );
$.mobile.pageContainer.trigger( "create" );

// Create checkboxes with plugin-driven options
$( '<fieldset data-role="controlgroup">
        <legend>Genre:</legend>
        <input type="checkbox" name="genre" id="c3" />
        <label for="c3">Action</label>...</fieldset>' )
        .insertAfter( "#create-cb2" );
$( '#c3' ).checkboxradio({ theme: "e" });
$( '#c4' ).checkboxradio({ theme: "e" });
$.mobile.pageContainer.trigger( "create" );
```

Slider

A slider is a common form control that allows users to select a value between a minimum and maximum range (see Figure 4–17).

Figure 4–17. *Slider*

For instance, in our example, we adjust the volume or brightness with a slider to adjust the range between a low and high setting. We can adjust the minimum and maximum boundaries of the slider and also set its default value. The user may adjust the slider by either sliding the control or by entering a value into the slider's corresponding text field. As shown in Listing 4–23, no additional markup is necessary for jQuery Mobile to enhance our slider. Any input element with type="range" will be automatically optimized.

Listing 4–23. *Slider (ch4/slider.html)*

```
<label for="volume">Volume:</label>
<input type="range" name="volume" id="volume" value="5" min="0" max="9"/>
```

A slider consists of two themeable components. There is the foreground component known as the slider and the background component known as the track. Each of these components can be themed separately. To theme the slider, add the `data-theme="a"` attribute to the input element. Additionally, to theme the track, add the `data-track-theme="a"` attribute to the input element:

```
<input type="range" name="brightness" id="brightness" min="0" max="10" data-theme="b"
data-track-theme="a" />
```

Dynamic Slider

The `slider` plugin is a multi-purpose widget that automatically enhances both sliders and switch controls. With this plugin we can dynamically create, enable, disable, and turn the switch control off and on (see Listing 4–24).

Listing 4–24. *Dynamic slider (ch4/dynamic-slider.html)*

```
// Create slider with markup-driven options
$( '<label for="s1">Brightness:</label>
 <input type="range" name="s1" id="s1" min="0" max="9" data-theme="d"/>')
        .insertAfter( "#element1" );
$( "#s1" ).slider();

// Create slider with plugin-driven options
$( '<label for="s1">Brightness:</label>
    <input type="range" name="s1" id="s1" min="0" max="10" />' )
        .insertAfter( "#create-s2" );
$( "#s1" ).slider({
        theme: "d",
        trackTheme: "a",
        disabled: false
});
```

Slider Options

The `slider` plugin has the following options:

`disabled` *boolean*
> **default:** `false`

> Disables the control. The `slider` plugin also has `enable` and `disable` methods to dynamically enable and disable the control.

> `$( "#element1" ).slider({ disabled: true });`

`initSelector` *CSS selector string*
> **default:** `"input[type='range'], :jqmData(type='range'), :jqmData(role='slider')"`

> The initSelector is used to define the selectors (element types, data roles, etc.) that trigger the automatic initialization of the `widget` plugin. For instance, all elements that are matched by the default selector will be enhanced by the

slider plugin. To override this selector, bind to the mobileinit event and update the selector as necessary:

```
$( document ).bind( "mobileinit", function(){
        $.mobile.slider.prototype.options.initSelector = "...";
});
```

theme *string*
> **default:** null. Inherited from parent.

> Sets the theme swatch color scheme for the slider. This is a letter from a to z that maps to the swatches included in your theme. By default, this will inherit the same swatch color as its parent container. This option is also exposed as a data attribute: data-theme=”a”.

> ```
> $( "#element1" ).slider({ theme: "a" });
> ```

trackTheme *string*
> **default:** null. Inherited from parent.

> Sets the theme swatch color scheme for the track the slider slides along. This is a letter from a to z that maps to the swatches included in your theme. By default, this will inherit the same swatch color as its parent container if not explicitly set. This option is also exposed as a data attribute: data-track-theme=”a”.

> ```
> $( "#element1" ).slider({ trackTheme: "a" });
> ```

Slider Methods

The slider plugin has the following methods:

enable: *enable a disabled slider or switch control.*
> ```
> $("#element1" ).slider( "enable" );
> ```

disable: *disable a slider or switch control.*
> ```
> $( "#element1" ).slider( "disable" );
> ```

refresh: *update a custom slider or switch control.*

> This updates the custom slider or switch to reflect the native element's value. For instance, we can dynamically update our switch or slider and call “refresh” to rebuild the control.

> ```
> // Set the switch to "on" and refresh it
> var switch = $( "#switch1" );
> switch[0].selectedIndex = 1;
> ```

```
switch.slider( "refresh" );

// Maximize the slider's volume control and refresh it
$( "#volume" ).val( 10 ).slider( "refresh" );
```

Slider Events

The slider plugin supports the following events:

create: triggered when a slider or switch control is created

> This event is triggered when a custom slider or switch control is created. It is not used to create a custom element.

```
$( '<select name="switch2" id="switch2">...</select>' )
        .insertAfter( "#element1" )
        .slider({
                create: function(event) {
                        console.log( "Creating new element..." );
                });
```

Switch Control

A switch control (see Figure 4–18) is commonly used to manage boolean on/off flags.

Figure 4–18. *Switch control*

For instance, switch controls are often the preferred means of allowing the user to manipulate application settings, due to their simplicity and ease of use. To flip the switch, the user can either tap the control or slide the switch. To create a switch control, add a select element with the data-role="slider" and two options to manage the on/off states (see Listing 4–25).

Listing 4–25. *Switch control (ch4/switch-control.html)*

```
<label for="alerts">Alerts:</label>
<select name="slider" id="alerts" data-role="slider">
        <option value="off">Off</option>
        <option value="on">On</option>
</select>
```

A switch control also consists of two themeable components. There is the foreground component known as the slider and the background component known as the track. Each of these components can be themed separately. To theme the slider, add the data-theme="a" attribute to the select element. Additionally, to theme the track, add the data-track-theme="a" attribute to the select element:

```
<select name="slider" data-theme="b" data-track-theme="c" data-role="slider">
```

```
        <option value="off">Off</option>
        <option value="on">On</option>
</select>
```

Dynamic Switch Control

As previously mentioned, the `slider` plugin is the widget that automatically enhances a switch control. With this plugin we can dynamically create, enable, disable, and turn the switch off and on (see Listing 4–26). The complete listing of the `slider` plugin was previously documented in the "Dynamic Slider" Section. The same API is reusable for both sliders and switches.

Listing 4–26. *Dynamic switch control (ch4/dynamic-switch-control.html)*

```
// Create switch with markup-driven options
$( '<select name="switch1" data-role="slider" data-theme="c"></select>' )
        .insertAfter( "#foo" )
        .slider();

// Create switch with plugin-driven options
$( '<select name="switch2" id="switch2">...</select>' )
        .insertAfter( "#switch1" )
        .slider({
                theme: "b",
                trackTheme: "c",
                disabled: false
        });
```

Native Form Elements

jQuery Mobile automatically enhances all form elements defined within your page. However, if you want to fall back to the native controls (see Figure 4–18) this can be configured globally or at the field level (see Listing 4–27).

Figure 4–19. *Native Form Elements*

Listing 4–27. *Native form elements (ch4/native.html)*

```
// Selectively choose which elements are native with data-role="none"
<label for="name">Text Input</label>
<input type="text" name="name" id="name" value="" data-role="none" />

<label for="slider2">Switch:</label>
<select name="slider2" id="slider2" data-role="none">
        <option value="off">Off</option>
        <option value="on">On</option>
</select>

// Globally configure native elements by selector
$(document).bind('mobileinit',function(){
        $.mobile.page.prototype.options.keepNative = "input, select";
});
```

To individually set a form field to display its native control, add the data-role="none"
attribute to its element. Alternatively, you can globally configure which form elements
should render natively by setting the keepNative selector when the mobileinit event
initializes. For instance, in Listing 4–27, we configured our selector to automatically
display all input and select elements in their native appearance. We will discuss how to
configure jQuery Mobile in greater depth in Chapter 8, "Configuring jQuery Mobile".

HTML5 provides several new input types to help collect date and time inputs. We now
have time, date, month, week, datetime, and datetime-local input types (see Listing 4-28).

Listing 4-28. *HTML5 dates (ch4/dates.html)*

```
<input type="time" name="time" />
<input type="datetime-local" name="dtl" />
<input type="date" name="date" />
<input type="month" name="month" />
<input type="week" name="week" />
<input type="datetime" name="dt" />
```

Support for these newer HTML5 input types is browser dependent (see
http://www.quirksmode.org/html5/inputs.html). The newer browsers that support
these features will display helpful date pickers (see Figure 4-20) and the unsupported
browsers will fallback to text inputs.

Figure 4-20. *HTML5 Dates*

Mobiscroll Date Picker

Mobiscroll[6] is an optimized date picker for touch screen devices. The Mobiscroll API is configurable,[7] which allows for the display of several date and time combinations (see Figure 4–21). Additionally, Mobiscroll is themable and can also be customized to display any data necessary (see Figure 4–22).

Figure 4–21. *Mobiscroll Date Picker*

Figure 4–22. *Mobiscroll with custom lists*

For example, we can update the MobiScroll options to create a customized movie search (see Listing 4–29). The Mobiscroll plugin is a flexible control that can be used for many different use cases.

---

[6] See http://code.google.com/p/mobiscroll/.

[7] See http://code.google.com/p/mobiscroll/wiki/Documentation.

Listing 4–29. *Mobiscroll (ch4/mobiscroll.html)*

```
// Import the Mobiscroll resources
<script type="text/javascript" src="jquery.scroller-1.0.2.js"></script>
<link type="text/css" rel="stylesheet" href="jquery.scroller-1.0.2.css"/>

// Display the default date picker (see Figure 4-21).
$( "#element1" ).scroller();

// Display a custom filter for a movie search (see Figure 4-22).
$( "#element2" ).scroller({
        setText: 'Search',
        theme: 'sense-ui',
        wheels: [{
                'Rating': { '5-star': '*****', '4-star': '****' ... },
                'Genre': { 'action': 'Action', 'comedy': 'Comedy', ...},
                'Screen': { '3d': '3D', 'imax': 'IMAX', 'wide': 'Wide' }
        }]
});
```

Summary

In this chapter, we reviewed every standard HTML form component and saw how jQuery Mobile automatically enhances each component to provide a unified user experience across all devices.

As we reviewed each component, we discussed its usage and identified common use cases it solves really well. We also reviewed the jQuery Mobile data attributes that are unique to each form element and saw code examples of how we can modify these attributes to configure and style our forms. Furthermore, we reviewed the plugins that are associated with each form component and saw how we can leverage the plugin API to dynamically create, enhance, and update our own components when users require a more dynamic experience.

Lastly, we explored the features of the Mobiscroll plugin and saw how it provides an elegant and flexible interface for date pickers, search filters, or custom lists.

In Chapter 5, our focus will shift from gathering user information to presenting user information. In particular, we will see the many ways we can style and configure lists of information.

List Views

Lists are a popular user interface component because they make the browsing experience very simple and efficient. Lists are also a very flexible component that can be styled in many ways and they adapt very well to different screen sizes. Whether we are browsing our mail, contacts, music, or settings, each of these apps display lists of information in slightly different styles. From basic lists that only include text to complex lists with graphics and detailed meta data, lists must be flexible enough to support many configurations. Fortunately, jQuery Mobile supports all of these list configurations and more. In this chapter we will explore the details of styling and configuring lists within jQuery Mobile. We will also see how to add search filters to our lists. Lastly, we will review the list view plugin API and see examples of how we can create and update lists dynamically.

List Basics

jQuery Mobile will automatically enhance any native HTML list (<ul> or <ol>) into a mobile optimized view when we add the data-role="list" attribute to our list element. The enhanced list will display edge-to-edge by default, and if our list items contain links, they will be displayed as touch-friendly buttons with a right-aligned arrow icon (see Figure 5–1 and the code snippet to produce it in Listing 5–1). By default, lists will be styled with the "c" swatch (gray) color. To apply an alternate theme, add the data-theme attribute to the list element or list items (<li>).

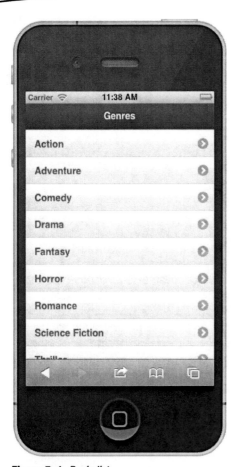

Figure 5–1. *Basic list*

Listing 5–1. *Basic list (ch5/list-basic.html)*

```
<ul data-role="listview" data-theme="c">
        <li><a href="#">Action</a></li>
        <li><a href="#">Adventure</a></li>
        <li><a href="#">Comedy</a></li>
</ul>
```

Inset Lists

An inset list will not appear edge-to-edge. Instead, an inset list is automatically wrapped inside a block with rounded corners and has margins set for additional spacing. To create an inset list, add the data-inset="true" attribute to the list element (see Figure 5–2 and 5–3, and the related code in Listing 5–2).

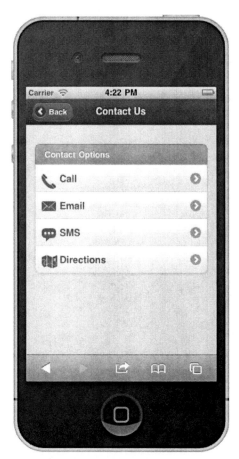

Figure 5–2. *Inset list (iOS)*

Figure 5–3. *Inset list (Windows Phone 7)*

Listing 5–2. *Inset list (ch5/list-inset.html)*

```
<ul data-role="listview" data-inset="true">
    <li data-role="list-divider">Contact Options</li>
    <li><a href="#"><img src="phone.png" class="ui-li-icon">Call</a></li>
    <li>...</li>
</ul>
```

List Dividers

A list divider can be used as a heading for a group of list items. For instance, if our app has a calendar listing, we may choose to group our calendar events by day (see Figure 5–4). List dividers can also be used as headers for inset lists. In our prior example, we also set the header of our inset list with a list divider (see Figure 5–2 and Listing 5–2).

To create a list divider, add the `data-role="list-divider"` attribute to any list item. The list divider's default text will appear left-aligned.

> **TIP:** In Figure 5–4, the list items contain both left-aligned and right-aligned text. To position text right-aligned, wrap it with an element that contains a class of `ui-li-aside` (see Listing 5–3).

By default, list dividers will be styled with the "`b`" swatch (light blue) color. To apply an alternate theme, add the `data-divider-theme="a"` attribute to the list element.

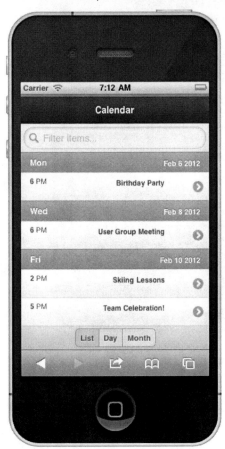

Figure 5–4. *List dividers*

Listing 5–3. *List dividers (ch5/list-dividers.html)*

```
<ul data-role="listview">
  <li data-role="list-divider" data-divider-theme="a">
    Mon <p class="ui-li-aside">Feb 6 2012</p>
  </li>
  <li>
    <a href="#">6 PM <span class="ui-li-aside">Birthday Party</span></a>
  </li>
</ul>
```

Lists with Thumbnails and Icons

We can add thumbnails to the left of our list item by adding an image inside a list item as the first child element (see Figure 5–5 and its related code in Listing 5–4). The framework will scale the image to 80 pixels square.

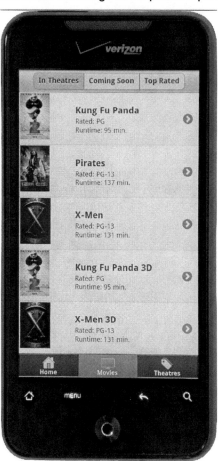

Figure 5–5. *List with thumbnails*

Listing 5–4. *List with thumbnails (ch5/list-thumbnails.html)*

```
<ul data-role="listview">
    <li>
        <a href="movies/kung-fu-panda.html">
            <img src="images/kungfupanda2.jpg" />
            <h3>Kung Fu Panda</h3>
            <p>Rated: PG</p>
            <p>Runtime: 95 min.</p>
        </a>
    </li>
    ...
</ul>
```

We can also use smaller icons instead of thumbnails. To use standard 16x16 pixel icons in list items, add the class of `ui-li-icon` to the image element (see Figure 5–6 and its related code in Listing 5–5).

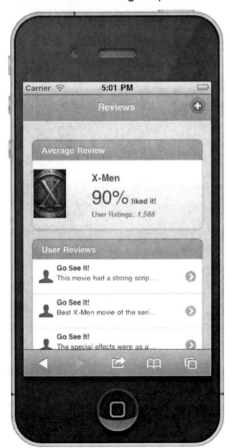

Figure 5–6. *List with icons*

Listing 5–5. *List with icons (ch5/list-icons.html)*

```
<ul data-role="listview" data-inset="true" data-theme="d">
      <li data-role="list-divider">User Reviews</li>
      <li>
          <a href="reviews/xmen/404.html">
              <img src="images/111-user.png" class="ui-li-icon">
              <p><strong>Go See It!</strong></p>
              <p>This movie had a strong script and ... </p>
          </a>
      </li>
      ...
</ul>
```

Split Button Lists

In situations where you need to support multiple actions per list item, we can create a split button list that has a primary and secondary button to choose from. For instance, we can modify our original movie listing example to support multiple actions. Our primary button will continue to show movie details and our new secondary button could be used to purchase tickets (see Figure 5–7).

Figure 5–7. *List with split buttons*

To create a split button, add a secondary link inside the list item and the framework will add a vertical line dividing the primary and secondary actions (see Listing 5–6).

Listing 5–6. *List with split buttons (ch5/list-split-buttons.html)*

```
<ul data-role="listview" data-split-icon="star" data-split-theme="d">
    <li>
        <a href="movies/kung-fu-panda.html">
            <img src="images/kungfupanda2.jpg" />
            <h3>Kung Fu Panda</h3>
            <p>Rated: PG</p>
            <p>Runtime: 95 min.</p>
        </a>
        <a href="tickets.html">Buy Tickets</a>
    </li>
    ...
</ul>
```

To set the icon for all secondary buttons, add the `data-split-icon` attribute to the list element and set its value to a standard (see Table 4-1) or custom icon. By default, the secondary button will be styled with the "b" swatch (light blue) color. To apply an alternate theme, add the data-split-theme attribute to the list element.

Numbered Lists

Numbered lists will be created when using an ordered list <ol> (see Figure 5–8 and its related code in Listing 5–7).

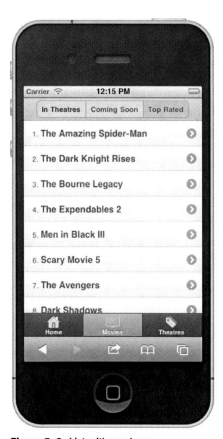

Figure 5–8. *List with numbers*

Listing 5–7. *List with numbers(ch5/list-numbered.html)*

```
<ol data-role="listview">
        <li><a href="spider-man.html">The Amazing Spider-Man</a></li>
        <li><a href="dark-knight.html">The Dark Knight Rises</a></li>
        ...
</ol>
```

By default the framework will add the numerical index to the left of each list item. These lists are useful when showing a listing of items that can be ranked sequentially. For instance, our "top rated" movies view is an ideal candidate for a numbered list because the sequence quickly associates which movies are rated highest.

Read-only Lists

List views can also display read-only views of data. The user interface looks very similar to our interactive views shown previously except the right arrow icon-only image has been removed and the font size and padding is slightly smaller. To create a read-only list, simply remove the anchor tags we used in our previous examples (see Figure 5–9 and its related code in Listing 5–8).

Figure 5–9. *List with read-only items*

Listing 5–8. *List with read-only items (ch5/list-readonly.html)*

```
<ul data-role="listview">
      <li>
            <img src="images/kungfupanda2.jpg" />
            <h3>Kung Fu Panda</h3>
            <p>Rated: PG</p>
            <p>Runtime: 95 min.</p>
      </li>
      ...
</ul>
```

List Badges (Count Bubbles)

A list badge or count bubble is a highlighted oval that typically indicates the number of new items that are available for viewing. For instance, badges are commonly used in mail applications to indicate how many unread email items you have. In our example, badges are used to indicate when a comment was added about a movie review (see Figure 5–10). A badge can be used to express any type of meta data.

Figure 5–10. *List with badges or count bubbles*

To create a badge, wrap the text of the badge with an element that contains a class of `ui-li-count`. By default, badges will be styled with the "c" swatch (gray) color. To apply an alternate theme, add the `data-count-theme` attribute to the list element (see Listing 5–9).

Listing 5–9. *List with badges or count bubbles (ch5/list-badges.html)*

```
<ul data-role="listview" data-inset="true" data-count-theme="e">
       <li data-role="list-divider">Comments</p></li>
       <li>
         <img src="../images/111-user.png" class="ui-li-icon">
```

```
        <p>Thanks for the review.  I'll check it out this weekend.</p>
        <span class="ui-li-count">1 day ago</span>
    </li>
</ul>
```

List Filtering with Search Bar

jQuery Mobile has a very convenient client-side search feature for filtering lists. To create a search bar, add the `data-filter="true"` attribute to the list. The framework will then append a search filter above the list and the default placeholder text will display the words, "Filter items…" (see Figure 5–11 and its related code in Listing 5–10).

Figure 5–11. List filtering (unfiltered)

Listing 5–10. *List filtering (ch5/list-filter.html)*

```
<ul data-role="listview" data-filter="true" data-filter-
        placeholder="Search...">
    <li data-role="list-divider">
        Mon <p class="ui-li-aside">Feb 6 2012</p>
    </li>
    <li>
        <a href="b-day.html">
            <p>6 PM <span class="ui-li-aside">Birthday Party</span></p>
        </a>
    </li>
</ul>
```

There are two options available for configuring the placeholder text:

1. You can configure the placeholder text by adding the `data-filter-placeholder`
    attribute on the list element (see Listing 5–10).

2. Or you may set the placeholder text globally as a jQuery Mobile configuration
    option by binding to the `mobileinit` event and setting the `filterPlaceholder`
    option to any custom placeholder value:

    ```
    $(document).bind('mobileinit',function(){
      $.mobile.listview.prototype.options.filterPlaceholder="Search..";
    });
    ```

We will discuss configuring jQuery Mobile in much greater detail in Chapter 8,
"Configuring jQuery Mobile".

As you begin entering text in the search filter, a client-side filter will only show items
matching the wildcard search (see Figure 5–12).

Figure 5–12. *List filtering (filtered)*

If you need to change the default search functionality there are two options for overriding the callback used for filtering:

First, you may update the search functionality globally as a jQuery Mobile configuration option by binding to the `mobileinit` event and setting the `filterCallback` option to any custom search function. For instance, here we set the callback to use a "starts with" search:

```
$(document).bind('mobileinit',function(){
    $.mobile.listview.prototype.options.filterCallback =
      function( text, searchValue ){
          // Use a "starts with" search
          return !(text.toLowerCase().indexOf( searchValue ) === 0);
      };
});
```

The callback function is provided two arguments, `text` and `searchValue`. The `text` argument contains the text of the list item and the `searchValue` argument contains the value of the search filter. The default behavior for the wildcard search is coded as:

```
              return text.toLowerCase().indexOf( searchValue ) === -1;
```

If the callback returns a truthy value for a list item it will be hidden from the search results.

Alternatively, we can also configure our search functionality dynamically after our list has been created. For instance, after our page loads we can apply our new search behavior for a specific list:

```
$("#calendar-list").listview('option', 'filterCallback',
        function( text, searchValue ) {
            // Use a "starts with" search
            return !(text.toLowerCase().indexOf( searchValue ) === 0);
        }
);
```

By default, the search box will inherit its theme from its parent container. To configure an alternate theme add the data-filter-theme attribute to the list element.

Dynamic Lists

The listview plugin is the widget that automatically enhances lists. We can leverage this plugin to dynamically create and update our lists. There are two options available for creating dynamic lists. You can create lists dynamically with a markup-driven approach or by explicitly setting the options on the listview plugin (see Listing 5–11).

Listing 5–11. *listview plugin examples (ch5/dynmic-lists.html)*

```
// Create list with markup-driven options
$( '<ul data-inset="true" id="list1">
        <li data-role="list-divider">Genres</li>
        <li><a href="#">Action</a></li>
        <li><a href="#">Comedy</a></li></ul>' )
    .insertAfter( "#list0" )
    .listview();

// Create list with plugin-driven options
$( '<ul><li data-role="list-divider">Genres</li>
        <li><a href="#">Action</a></li>
        <li><a href="#">Comedy</a></li></ul>' )
    .insertAfter( "#list1" )
    .listview({
        theme: "d",
        dividerTheme: "a",
        inset: true,
    });

// Add a new item to an existing list
$( "#list1" )
        .append('<li><a href="#">Drama</a></li>')
        .listview("refresh");
```

List Options

The `listview` plugin has the following options:

countTheme *string*
>   **default:** `"c"`
>
>   Sets the theme swatch color scheme for the badges or count bubbles. This is a letter from a-z that maps to the swatches included in your theme. This option is also exposed as a data attribute: data-count-theme="a".
>
>   `$( "#list1" ).listview({ countTheme: "a" });`

dividerTheme *string*
>   **default:** `"b"`
>
>   Sets the theme swatch color scheme for the list dividers. This is a letter from a-z that maps to the swatches included in your theme. This option is also exposed as a data attribute: data-divider-theme="a".
>
>   `$( "#list1" ).listview({ dividerTheme: "a" });`

initSelector CSS selector string
>   **default:** `":jqmData(role='listview')"`
>
>   The initSelector is used to define the selectors (element types, data roles, etc.) that are used to trigger the automatic initialization of the widget plugin. For instance, all elements that are matched by the default selector will be enhanced by the `listview` plugin. To override this selector, bind to the `mobileinit` event and update the selector as needed:

```
$( document ).bind( "mobileinit", function(){
    $.mobile.listview.prototype.options.initSelector = "...";
} );
```

inset *boolean*
>   **default:** `false`
>
>   An inset list will be created when this option is set to true. By default, a basic list will be created. This option is also exposed as a data attribute: data-inset="true".
>
>   `$( "#list1" ).listview({ inset: true });`

splitIcon *string*
>   **default:** `"arrow-r"`
>
>   Sets the icon for the secondary button when building a split button list. This option is also exposed as a data attribute: data-split-icon="star".
>
>   `$( "#list1" ).listview({ splitIcon: "star" });`

splitTheme *string*
>   **default:** `"b"`
>
>   Sets the theme swatch color scheme for the secondary button when creating a split button list. This is a letter from a-z that maps to the swatches included in your theme. This option is also exposed as a data attribute: data-split-theme="a".

```
$( "#list1" ).listview({ splitTheme: "a" });
```
theme *string*
> **default:** "c"

Sets the theme swatch color scheme for the list. This is a letter from a-z that maps to the swatches included in your theme. This option is also exposed as a data attribute: data-theme="a".

```
$( "#list1" ).listview({ theme: "a" });
```

List Methods

The listview plugin has the following methods:

refresh: updates the custom list.

This updates the custom list to reflect the native list element's value. For instance, if we add a new item to an existing list, we must call "refresh" to rebuild the list items:

```
// Add an item to an existing list and refresh the list item
$( "#list1" )
        .append('<li><a href="#">Drama</a></li>')
        .listview("refresh");

// Add list items to a new list and refresh the entire list
var markup = '<li>item 1</li><li>item 2</li>';
$( "#list2" )
        .append(markup)
        .listview( "refresh", true );
```

List Events

The listview plugin supports the following events:

create: triggered when a list is created

This event is triggered when a custom list is created. It is not used to create a custom list.

```
$( '<ul><li data-role="list-divider">Genres</li>
    <li><a href="#">Comedy</a></li></ul>' )
    .insertAfter( "#list1" )
    .listview({
        inset: true,
        create: function(event) {
                console.log( "Creating list..." );
        }
    });
```

Summary

In this chapter, we reviewed the very popular list view component. List views are commonly used because they make the browsing experience very simple and efficient. jQuery Mobile lists can be styled and configured in many unique ways. From basic lists to lists with images, split buttons, dividers, or badges, we have many configuration options to choose from.

We also saw how easy it was to add a search filter to our lists and saw examples of how we could override the default search if necessary.

Lastly, we reviewed the `listview` plugin API and saw examples of how to dynamically create and update lists to provide our users with a more interactive experience.

In our next chapter, we will explore another popular user interface component, jQuery Mobile's flexible grid layout. We will see how we can use grids to create responsive designs and we will also look at enhancing our user interface with CSS gradients.

Chapter 6

Formatting Content with Grids and CSS Gradients

Mobile applications often use grids for content that needs to be flexible and grouped into sections. jQuery Mobile's responsive grid is a useful solution for designs that require this behavior. In this chapter, we will review the basics of the jQuery Mobile grid component and show several examples of how we can style icons, graphics, and text in our grids. We will also create collapsible content blocks and discuss the advantage they provide when compared to an inline page structure. Lastly, we will add a bit of polish to our designs with CSS gradients and discuss the advantages that CSS gradients provide in regards to performance and progressive enhancement.

Grid Layouts

jQuery Mobile's grids are configurable to support layouts in the range of two to five columns. From an HTML perspective, grids are `div` elements that are configured with CSS attributes. The Grid is flexible and will consume the entire width of your display. The grids do not contain borders, padding, or margins so they will not interfere with the styles of elements contained within them. Before we look at an example let's review the standard grid template.

Grid Template

The grid template may be a helpful reference when creating multi-column grids (see Listing 6–1).

Listing 6–1. *Grid Template*

```
<div data-role="content">

  <!-- Grid container -->
  <div class="<grid-css-attribute>">
```

```
<!-- Blocks -->
<div class="<block-css-attribute>">Block A</div>
<div class="<block-css-attribute>">Block B</div>

    </div>
</div>
```

When creating a grid, you will be required to create the outer grid container with two or more inner blocks:

1. Grid container: The grid container requires the CSS attribute `ui-grid-*` to configure the number of columns in the grid (see Table 6–1). For instance, to create a two-column grid we would set our grid CSS attribute to `ui-grid-a`.

Table 6–1. *Grid CSS Attribute Reference*

Number of Columns	Grid CSS Attribute
2	ui-grid-a
3	ui-grid-b
4	ui-grid-c
5	ui-grid-d

2. Blocks: The blocks are contained within the grid. The blocks require the CSS attribute `ui-block-*` to identify its column position (see Table 6–2). For instance, if we had a two-column grid, our first block would be styled with CSS attribute `ui-block-a` and the second block would be styled with CSS attribute `ui-block-b`.

Table 6–2. *Block CSS Attribute Reference*

Column	Block CSS Attribute
1.	ui-block-a
2.	ui-block-b
3.	ui-block-c
4.	ui-block-d
5.	ui-block-e

Two-Column Grid

A two-column (50%, 50%) grid is shown in Figure 6–1 with its related code in Listing 6–2.

Figure 6–1. *Two-column grid*

Listing 6–2. *Two-column grid (ch6/grid-2col.html)*

```
<div data-role="content">
  <div class="ui-grid-a">
    <div class="ui-block-a">Block A</div>
    <div class="ui-block-b">Block B</div>
  </div>
</div>
```

The outer grid is configured with the CSS grid attribute of ui-grid-a. Next, we add two internal blocks. The first block is assigned a CSS value of ui-block-a and the second column is assigned a CSS value of ui-block-b. As shown in Figure 6–1, the columns are equally spaced, borderless, and the text within each block will wrap when necessary. As an added bonus, the grids within jQuery Mobile are flexible and will render responsively across different display sizes (see Figure 6–2).

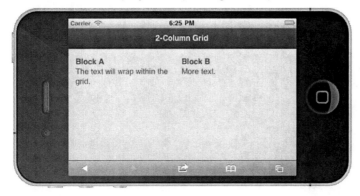

Figure 6–2. *Two-column grid (landscape)*

Three-Column Grid with CSS Enhancements

A three-column (33%, 33%, 33%) grid is shown in Figure 6–3 with its related code in Listing 6–3.

Figure 6–3. *Three-column grid with CSS enhancements*

Listing 6–3. *Three-column grid (ch6/grid-3col.html)*

```
<div data-role="content">
    <div class="ui-grid-b">
        <div class="ui-block-a">
            <div class="ui-bar ui-bar-e" style="height:100px">Block A</div>
        </div>
        <div class="ui-block-b">
            <div class="ui-bar ui-bar-e" style="height:100px">Block B</div>
        </div>
        <div class="ui-block-c">
            <div class="ui-bar ui-bar-e" style="height:100px">Block C</div>
        </div>
    </div>
</div>
```

It closely resembles the two-column example we saw earlier except the CSS attribute for the grid is configured to support three columns (ui-grid-b) and we have added an additional block for the third column (ui-block-c). We also styled the blocks with themeable classes which can be added to any element including grids. In the example, we added ui-bar to apply css padding and added ui-bar-e to apply the background gradient and font styling for the "e" toolbar theme swatch. You may style your blocks with any toolbar theme (ui-bar-*) in the range a through e. Lastly, to create consistent block heights we also styled our height inline (style="height:100px"). Visually, these enhancements have styled our grid with a linear background gradient and our blocks are now segregated with borders.

Four-Column Grid with App Icons

A four-column (25%, 25%, 25%, 25%) grid is shown in Figure 6–4 with its related code in Listing 6–4.

Listing 6–4. *Four-column grid (ch6/grid-4col.html)*

```
<div data-role="content">
    <div class="ui-grid-c" style="text-align: center;">
        <div class="ui-block-a"><img src="images/cloud-default.png"></div>
        <div class="ui-block-b"><img src="images/cloud-bright.png"></div>
        <div class="ui-block-c"><img src="images/cloud-ripple.png"></div>
        <div class="ui-block-d"><img src="images/cloud-sparkle.png"></div>
    </div>
</div>
```

It is similar to the three-column example except the CSS attribute for the grid is configured to support four columns (ui-grid-c) and we have added an additional block for the fourth column (ui-block-d). Additionally, we centered the app icons within the grid for balance and consistency (style="text-align:center;"). Visually, the grid has evenly spaced app icons that closely resembles an application springboard.

Figure 6–4. *Four-column grid with app icons*

Five-Column Grid with Emoji Icons

A five-column (20%, 20%, 20%, 20%, 20%) grid is shown in Figure 6–5 with its related code in Listing 6–5.

Listing 6–5. *Five-column grid (ch6/grid-5col.html)*

```
<div data-role="content">
  <div class="ui-grid-d" style="text-align: center;">
    <div class="ui-block-a">&#xe21c;</div>
    <div class="ui-block-b">&#xe21d;</div>
    <div class="ui-block-c">&#xe21e;</div>
    <div class="ui-block-d">&#xe21f;</div>
    <div class="ui-block-e">&#xe220;</div>
  </div>
</div>
```

The example closely resembles the four-column grid we saw previously except the CSS attribute for the grid is configured to support five columns (ui-grid-d) and we have added an additional block for the fifth column (ui-block-e). Each block contains a unique Emoji icon.[1]

Figure 6–5. *Five-column grid*

NOTE: Emoji icons are a performant alternative to images because they consume zero HTTP requests and their payload is only a few characters of text. Unfortunately, Emoji icons are currently only supported in iOS.

---

[1] See http://pukupi.com/post/1964.

Multi-Row Grid

Thus far we have only seen grids with a single row. To add an additional row, simply repeat the block pattern of the first row for each consecutive row (see Figure 6–6 and its related code in Listing 6–6). The resulting grid contains five columns and three rows. The columns are evenly spaced and you may manually adjust the row height at the block component.

Figure 6–6. *Multi-row grid*

Listing 6–6. *Multi-row grid (ch6/grid-multi-row.html)*

```
<div data-role="content">
  <div class="ui-grid-d" style="text-align: center;">

    <!-- First row -->
    <div class="ui-block-a">&#xe21c;</div>
    <div class="ui-block-b">&#xe21d;</div>
    <div class="ui-block-c">&#xe21e;</div>
    <div class="ui-block-d">&#xe21f;</div>
    <div class="ui-block-e">&#xe220;</div>
```

```
<!-- Second row -->
<div class="ui-block-a">&#xe002;</div>
<div class="ui-block-b">&#xe005;</div>
<div class="ui-block-c">&#xe51a;</div>
<div class="ui-block-d">&#xe515;</div>
<div class="ui-block-e">&#xe152;</div>

</div>
</div>
```

Uneven Grids

So far, every grid example shown had evenly spaced columns because jQuery Mobile
will space all columns equally by default. However, if you need to customize the column
dimensions we can adjust the widths in CSS. For instance, we can modify the default
widths in our 2-column grid to a 25/75% grid by setting the custom width of each block
(see Figure 6–7 and its related code in Listing 6–7). As a result, our grids can be
modified to support a wide range of alternate dimensions.

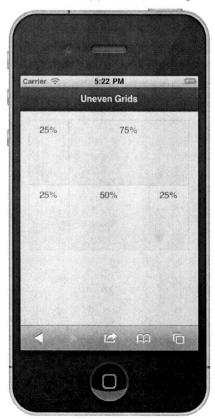

Figure 6–7. *Uneven grid*

Listing 6–7. *Uneven grid (ch6/grid-uneven.html)*

```
<style>
        /* Set 2-column grid to 25/75% */
        .ui-grid-a .ui-block-a {
            width: 25%;
        }
        .ui-grid-a .ui-block-b {
            width: 75%;
        }

        /* Set 3-column grid to 25/50/25% */
        .ui-grid-b .ui-block-a {
            width: 25%;
        }
        .ui-grid-b .ui-block-b {
            width: 50%;
        }
        .ui-grid-b .ui-block-c {
            width: 25%;
        }
</style>
```

Springboard

A springboard is an ideal candidate for applying our grid layout. In the examples below we will see two types of springboards. First, we will see a springboard styled with app icons (see Figure 6–8) and secondly we will see a springboard styled with Glyphish icons (see Figure 6–9).

Figure 6–8. *Springboard with app icons*

Figure 6–9. *Springboard with Glyphish icons*

Are you up for a springboard challenge? If you are, I encourage you to create a springboard that closely resembles one of the two figures. Both examples are configured identically from a grid perspective. However, the Springboard with Glyphish icons (see Listing 6–9) is styled slightly different than the Springboard with app icons (see Listing 6–8) to adjust for its uneven icon heights.

Listing 6–8. *Springboard with app icons (ch6/springboard1.html)*

```
<div class="ui-grid-a">
  <div class="ui-block-a">
    <a href="#">
      <img src="images/cloud.png">
      <span class="icon-label">App A</span>
    </a>
  </div>
  ...
</div>

<style>
  /* center icons */
```

```
  .ui-grid-a { text-align: center; }

  /* set row height */
  .ui-block-a, .ui-block-b { height: 100px; }

  /* set label color and size */
  .icon-label { color: #000; display: block; font-size:12px; }

  a:link, a:visited, a:hover, a:active { text-decoration:none; }
</style>
```

Listing 6–9. *Springboard with Glyphish icons (ch6/springboard2.html)*

```
<div class="ui-grid-a">
  <div class="ui-block-a">
    <div class="icon-springboard">
      <a href="#">
        <img src="images/45-movie-1-lg.png" alt="Now Playing">
        <span class="icon-label">Now Playing</span>
      </a>
    </div>
  </div>
  …
</div>

<style>
  /* center icons */
  .ui-grid-a { text-align: center; }

  /* set row height */
  .ui-block-a, .ui-block-b { height: 100px; position: relative; }

  /* set label size and color */
  .icon-label { color: #FFF; display: block; font-size:12px; }

  /* bottom align icons to adjust for uneven icon heights */
  .icon-springboard { position: absolute; bottom: 0; width: 100%; }

  a:link, a:visited, a:hover, a:active { text-decoration:none; }
</style>
```

Collapsible Content Blocks

Do you ever find yourself scrolling repeatedly to view the contents of an entire mobile page? Although this may be a good workout for your finger it can be a cumbersome user experience when users must scroll repeatedly. If you are looking for a more usable alternative, you may want to consider grouping your content into collapsible content blocks.

> **TIP:** Collapsible content blocks have several advantages when compared to an inline page structure. First, we can collapse content into segmented groups to make them all visible within a single view (see Figure 6–10). And secondly, our users will be more efficient because we have eliminated scrolling from the user experience.

Figure 6–10. *Content block (all blocks collapsed)*

The markup required to create a collapsible content block is shown in Listing 6–10.

Listing 6–10. *Collapsible content block (ch6/collapsible-block.html)*

```
<div data-role="content">

  <div data-role="collapsible" data-collapsed="true" data-theme="a" data-content-theme="b">
    <h3>Wireless</h3>
    <ul data-role="listview" data-inset="true">
      <li><a href="#">&#xe117; Notifications</a></li>
      <li><a href="#">&#xe01d; Location Services</a></li>
    </ul>
```

```
    </div>

    <div data-role="collapsible" data-theme="a" data-content-theme="b">
      <h3>Applications</h3>
      <ul data-role="listview" data-inset="true">
        <li><a href="#">&#xe001; Faceoff</a></li>
        <li><a href="#">&#xe428; LinkedOut</a></li>
        <li><a href="#">&#xe03d; Netflicks</a></li>
      </ul>
    </div>

</div>
```

There are two required elements for creating a collapsible block:

1. Create a container and add the data-role="collapsible" attribute. Optionally, you may configure the container to be collapsed or expanded by adding the data-collapsed attribute. By default, a collapsible section will be shown expanded (data-collapsed="false"). To initially show the section as a collapsed block, add data-collapsed="true" to the container. For instance, if we launch the code in Listing 6–10, the initial view will appear as Figure 6–11. In the code listing, we have explicitly collapsed all content blocks except for the "Applications" section which will expand by default.

Figure 6–11. *Content block (one block expanded)*

2. Within the container, add any header element (H1-H6). The framework will style the header to look like a clickable button with a left-aligned plus or minus icon to indicate it's expandable.

After the header, you may add any HTML markup to the collapsible block. The framework will wrap this markup within the container that will expand or collapse when the heading is tapped. You may theme the collapsible block and its associated button separately by adding the data-theme and data-content-theme attributes to the collapsible container (see Listing 6-10).

> **NOTE:** A collapsible block allows you to have many blocks expanded or collapsed at once (see Figure 6–12). In our next section, we will see this is not allowed when working with collapsible sets.

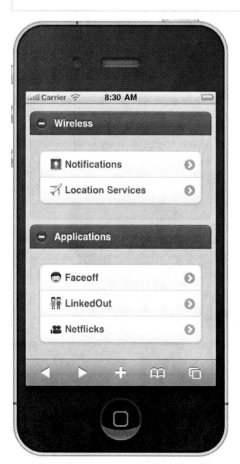

Figure 6–12. *Content block (all blocks expanded)*

Collapsible Sets

Collapsible sets (see Figure 6–13) are similar to collapsible blocks except their collapsible sections are visually grouped together and only one section may be expanded at a time which gives the collapsible set the appearance of an accordion (see Figure 6–14).

Figure 6–13. *Content set (collapsed)*

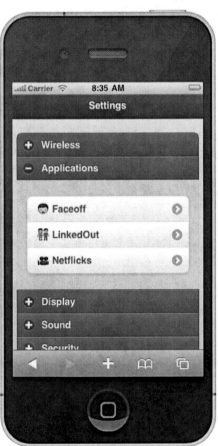

Figure 6–14. *Content set (expanded)*

When opening a new section within the set, any section that was previously opened will collapse automatically.

The markup for a collapsible set is identical to the markup we saw previously when building a collapsible block. However, to create the accordion-style behavior and grouping we need to add a parent wrapper with a `data-role="collapsible-set"` as shown in Listing 6–11. Again, you may theme the collapsible section and its associated button separately by adding the `data-theme` and `data-content-theme` attributes to the collapsible set.

Listing 6–11. *Collapsible set (ch6/collapsible-set.html)*

```
<div data-role="content">

  <div data-role="collapsible-set" data-theme="a" data-content-theme="b">
    <div data-role="collapsible" data-collapsed="true">
      <h3>Wireless</h3>
      <ul data-role="listview" data-inset="true">
        <li><a href="#">&#xe117; Notifications</a></li>
        <li><a href="#">&#xe01d; Location Services</a></li>
      </ul>
    </div>

    <div data-role="collapsible">
      <h3>Applications</h3>
      <ul data-role="listview" data-inset="true">
        <li><a href="#">&#xe001; Faceoff</a></li>
        <li><a href="#">&#xe428; LinkedOut</a></li>
        <li><a href="#">&#xe03d; Netflicks</a></li>
      </ul>
    </div>
    ...
  </div><!-- /collapsible-set -->
</div>
```

Styling with CSS Gradients

Looking to add a bit of polish to your mobile UI? Try using CSS gradients where you would typically use background images. CSS gradients offer a performant alternative to images, they work extremely well within flexible layouts, and they gracefully degrade in unsupported browsers. For example, we can take an original springboard (see Figure 6–15) and transform it into a much more elegant display (see Figures 6–16 and 6–17) with the addition of gradients.

Figure 6–15. *Springboard without CSS gradients*   **Figure 6–16.** *Springboard with CSS gradients (iOS))*

Figure 6–17. *Springboard with CSS gradients (Android)*

Gradients can be used anywhere background images are used. For example, they are most commonly used to style backgrounds of your header, content, and buttons. Furthermore, there are two types of CSS gradients: linear and radial. Linear gradients are the simpler of the two and if you are not too familiar with their syntax there are CSS gradient generators[2] available to help you get started. The CSS to generate the linear gradient of our background is shown in Listing 6–12.

Listing 6–12. *Background gradient*

```
.background-gradient {
  background-image: -webkit-gradient(
    linear, left bottom, left top,
    color-stop(0.22, rgb(92,92,92)),
    color-stop(0.57, rgb(158,153,158)),
```

---

[2] See http://www.westciv.com/tools/gradients/ or http://gradients.glrzad.com/.

```
      color-stop(0.84, rgb(92,92,92))
  );
}
```

```
<!-- Set the gradient on the page -->
<div data-role="page" class="background-gradient">
```

While this CSS gradient is targeted for the most popular WebKit layout engine (see Figure 6–18) you may add support for additional browsers by including their vendor-specific prefix.

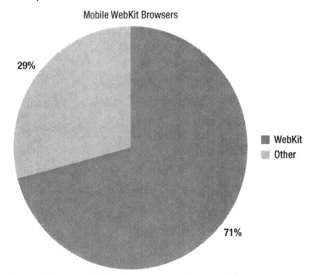

http://en.wikipedia.org/wiki/Mobile_browser#Popular_mobile_browsers

Figure 6–18. *Webkit usage*

For example, to render our gradient on Mozilla browsers we would add the -moz- vendor prefixed version (see Listing 6–13).

Listing 6–13. *Background gradient with Mozilla support*

```
.background-gradient {
  background-image: -webkit-gradient(
    linear, left bottom, left top,
    color-stop(0.22, rgb(92,92,92)),
    color-stop(0.57, rgb(158,153,158)),
    color-stop(0.84, rgb(92,92,92))
  );
  background-image: -moz-linear-gradient(
    90deg,
    rgb(92,92,92),
    rgb(158,153,158),
    rgb(92,92,92));
}
```

The gradient for the header is actually an overlay of three separate gradients. Including one linear gradient and two radial gradients. A radial gradient creates a circular gradient effect. The code to create our header gradient is shown in Listing 6–14.

Listing 6–14. *Springboard gradient*

```
.header-gradient {
  background-image:
    -webkit-gradient(
      linear, left top, left bottom,
      from( rgba( 068,213,254,0 )),
      color-stop(.43, rgba( 068,213,254,0 )),
      to( rgba( 068,213,254,1 ))),
    -webkit-gradient( radial,
      50% 700, 690,
      50% 700, 689,
      from( rgba( 049,123,220,0 )),
      to( rgba( 049,123,220,1 ))),
    -webkit-gradient(
      radial,
      20 -43, 60,
      20 -43, 40,
      from( rgba( 125,170,231,1 )),
      to( rgba( 230,238,250,1 )));
}

<!-- Set the gradient on the header -->
<div data-role="header" class="header-gradient">
```

Summary

In this chapter, we reviewed the usefulness of jQuery Mobile's grid-based design and saw how quickly we can style content within our grid template (see Listing 6–1). The jQuery Mobile grid is an ideal solution for content that needs to be responsive and grouped into sections. Our grids can contain any content and we saw several examples of grids styled with text, icons, and graphics.

We also reviewed collapsible content blocks and discussed their advantages when compared to an inline page structure. Collapsible blocks can be an effective usability pattern because they help display all content within a single view and they help eliminate scrolling from the user experience. As a result, the users' experience with the app will be more efficient.

Lastly, we saw how to polish our designs with CSS gradients. CSS gradients are a performant alternative to images, they work extremely well within flexible layouts, and they progressively degrade in unsupported browsers.

In Chapter 7 we will continue down our path of layout design and take a closer look at jQuery Mobile's theming framework.

Creating Themable Designs

jQuery Mobile has a built-in theming framework that allows designers to quickly customize or re-style their user interface. The theming framework takes advantage of many CSS3 features, which helps build more elegant and responsive designs. For instance, by leveraging CSS3, the theming framework is able to apply rounded corners, shadows, and gradients without having to rely upon images. This is a performance advantage in that the framework can provide a more attractive interface without the overhead of extra HTTP requests. Essentially, we have a lightweight theming framework that renders a unified design across all browsers.

In this chapter, we will discuss the basics of the theming framework and review the default themes that are included in jQuery Mobile. We will also explore the three ways themes can be assigned to components. While all components can have their theme explicitly set with the data-theme attribute, most components also have default themes and may also inherit their theme from a parent container. We will discuss their advantages, look at examples of each, and discuss the precedence in which themes are applied to components.

Lastly, we will see how we can create our own custom themes. Creating custom themes will be necessary if you need to create richer designs or if you need to create a design that closely matches your corporate branding. There are two options available for creating custom themes and we will look at step-by-step examples of each. The first option is the manual approach, which gives designers complete control of their layout; the second is to use ThemeRoller[1], a web-based tool that automates the process of creating new themes.

---

[1] See http://jquerymobile.com/themeroller

Theme Basics

In many examples we have already seen how we can use the data-theme attribute to apply alternate themes to our page containers (page, header, content, footer) and form elements. For instance, we can take an un-themed page (see Figure 7–1) and re-style it with a different header and list theme (see Figure 7–2) with the simple addition of data-theme attributes (see Listing 7–1).

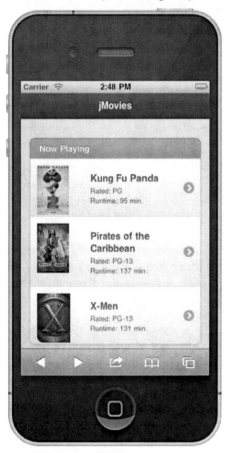

Figure 7–1. *List with default theme*

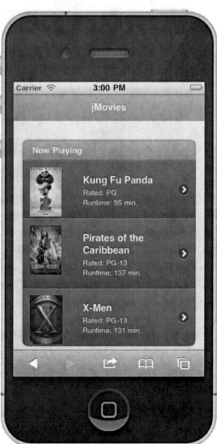

Figure 7–2. *List with alternate theme*

Listing 7–1. *data-theme attribute (ch7/theme-list2.html)*

```
<div data-role="page">
    <div data-role="header" data-theme="b">
        <h1>jMovies</h1>
    </div>

    <div data-role="content">
        <ul data-role="listview" data-inset="true" data-theme="a">
            <li data-role="list-divider">Now Playing</li>
```

```
            </ul>
        </div>
</div>
```

Themes and Swatches

The jQuery Mobile CSS file is always the first asset we import in the `head` element (see
Listing 7–2). This file contains the default structure and theming for jQuery Mobile
applications. Take a moment and explore the contents of this file with your favorite
editor.

Listing 7–2. *jQuery Mobile CSS import*

```
<head>
    <link rel="stylesheet" type="text/css" href="jquery.mobile-min.css" />
    <script type="text/javascript" src="jquery-min.js"></script>
    <script type="text/javascript" src="jquery.mobile-min.js"></script>
</head>
```

The jQuery Mobile CSS document is broken into two sections: a theme section and a
structure section.

▨ Theme – The top half of the document contains the default theme
settings. The theme settings manage the visual styling (backgrounds,
borders, color, font, shadows) for all components. When setting the
`data-theme` attribute, we are able to choose from five different options
(a, b, c, d, e). These letters (a-e) are technically referred to as
swatches. As you were reviewing the jQuery Mobile CSS file you may
have noticed that the first swatch to appear within the CSS file was
swatch "a" (see Listing 7–3).

Listing 7–3. *jQuery Mobile CSS swatch "a" (partial listing)*

```
/* A
--------------------------------------------------------------------*/
.ui-bar-a {
        border: 1px solid           #2A2A2A;
        background:                 #111111;
        color:                            #ffffff;
        font-weight: bold;
        text-shadow: 0 -1px 1px #000000;
        ...
        background-image:           linear-gradient(top, #3c3c3c, #111);
}
.ui-body-a {
        border: 1px solid           #2A2A2A;
        background:                 #222222;
        color:              #fff;
        text-shadow: 0 1px 0    #000;
        font-weight: normal;
        background-image:           linear-gradient(top, #666, #222);
}
...
```

The theme section is broken down into the following sub-sections:

■ Swatches – By default, jQuery Mobile has five swatches to choose from (a, b, c, d, e) and you may add as many unique swatches as necessary. Swatches allow us to configure unique backgrounds, borders, colors, fonts, and shadows for our components. For simplicity, the naming convention for new swatches is letter based (a-z). However, there is no limitation to the length of the swatch names. We will see examples of creating our own custom swatches later in the chapter.

■ Global theme settings – Global theme settings are configured after the swatches. These settings add visual styling enhancements to buttons, such as rounded corners, icons, overlays, and shadows. Since these settings are global, they will be inherited by all swatch configurations (see Listing 7–4).

Listing 7–4. *jQuery Mobile global theme styling (partial listing)*

```
/* Active class used as the "on" state across all themes
--------------------------------------------------------------------*/
.ui-btn-active {
        border: 1px solid            #155678;
        background:                  #4596ce;
        font-weight: bold;
        color:              #fff;
        cursor: pointer;
        text-shadow: 0 -1px 1px #145072;
        text-decoration: none;
        ...
}
```

■ Structure – The latter half of the jQuery Mobile CSS file contains structure styling that primarily includes positioning, padding, margin, height, and width settings (see Listing 7–5).

Listing 7–5. *jQuery Mobile structure styling (partial listing)*

```
/* some unsets - more probably needed */
.ui-mobile, .ui-mobile body { height: 100%; }
.ui-mobile fieldset, .ui-page { padding: 0; margin: 0; }
.ui-mobile a img, .ui-mobile fieldset { border: 0; }


...

.ui-checkbox, .ui-radio {
        position:relative;  margin: .2em 0 .5em; z-index: 1;
}
.ui-checkbox .ui-btn, .ui-radio .ui-btn {
        margin: 0; text-align: left; z-index: 2;
}
```

Now that we have been introduced to the main CSS file for jQuery Mobile, let's take a closer look at the five swatches that are included with jQuery Mobile and see how they appear across several different components (see Figure 7–3–7–6).

Figure 7–3. *Grid swatches*

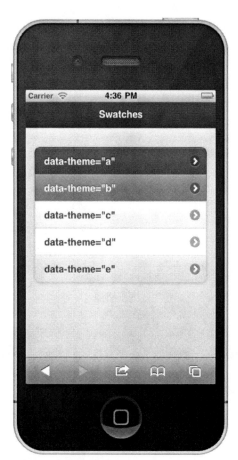

Figure 7–4. *List swatches*

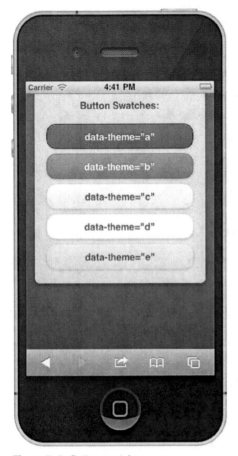

Figure 7–5. *Button swatches*

Figure 7–6. *Form field swatches*

To keep the styling of the swatches consistent across all components, the following visual priority conventions are used for each swatch:

- "a" - (black) highest level of visual priority.
- "b" - (blue) is secondary level.
- "c" - (gray) baseline.
- "d" - (white/gray) an alternate secondary level.
- "e" - (yellow) an accent color.

Theme Defaults

If you do not add data-theme attributes to a page, jQuery Mobile will apply default themes for all page containers and form elements (see Table 7–1).

Table 7–1. *Themes by component*

Component	Default Theme	Inherit's Parent Theme?	Example
Button	Inherited from parent	Yes	Listing 4-8
Checkbox	Inherited from parent	Yes	Listing 4-20
Content	data-theme="c"	Yes	Listing 7–6
Dialog	data-theme="a"	Yes	Listing 2-6
Grid	None	Yes	Listing 6-3
Footer	data-theme="a"	No	Listing 7–6
Header	data-theme="a"	No	Listing 7–6
List view	data-theme="c"	No	Listing 5-1
List badge	data-theme="c"	No	Listing 5-9
List divider	data-theme="b"	No	Listing 5-3
List item	data-theme="c"	Yes (from list only)	Listing 7–6
List split button	data-theme="b"	No	Listing 5-6
Page	data-theme="c"	No	Listing 7–5
Radio button	Inherited from parent	Yes	Listing 4-18
Select	Inherited from parent	Yes	Listing 4-16
Slider	Inherited from parent	Yes	Listing 4-22
Switch	Inherited from parent	Yes	Listing 4-24
Text input	Inherited from parent	Yes	Listing 4-13

For instance, if we create a basic jQuery Mobile page without explicitly setting its
themes, our elements will fall back to their default theme or inherit the theme of their

parent container. In Figure 7–7, default themes were applied to the page, header, footer, content, and list elements, whereas the form elements inherit their themes.

Figure 7–7. *Page with default and inherited themes*

By referencing our "Default themes by component" Table (see Table 7–1), we can determine what defaults will be applied for each component. Let's take a closer look at the content and button components. By default, the content component will have data-theme="c" applied. However, the button component does not have a default theme, so it will inherit its default theme from its parent container. In Listing 7–6, the button's parent is the content container; as a result, the button will inherit theme "c". Moreover, if the button was within the head container, it would inherit the head container's theme.

Listing 7–6. *Page with default themes (ch7/theme-defaults.html)*

```
<div data-role="page">
    <div data-role="header">
        <h1>default = "a"</h1>
    </div>

    <div data-role="content">
      default = "c"
```

```
    <ul data-role="listview" data-inset="true">
        <li data-role="list-divider">default = "b"</li>
        <li>default = "c"</li>
        <li>default = "c"</li>
    </ul>

    <form id="test" id="test" action="#" method="post">
        <p>
            <label for="text">inherits "c":</label>
            <input type="text" name="text" id="text" value="" />
        </p>
        <p>
            <label for="sound">inherits "c":</label>
            <select name="slider" id="sound" data-role="slider">
                <option value="off">Off</option>
                <option value="on">On</option>
            </select>
        </p>

        <a href="#" data-role="button">Button (inherits "c")</a>
    </form>
</div>

<div data-role="footer" data-position="fixed">
    <h3>default = "a"</h3>
</div>
</div>
```

Theme Inheritance

Components can also inherit the themes of their parent containers. Theme inheritance is beneficial in two regards. First, it makes the styling process more efficient for designers because we can rapidly set a theme at a high level (page container) and that theme will cascade down to all sub-components, saving valuable time. Secondly, it keeps components styled consistently across the entire application. For instance, in Listing 7–7, we have styled our page container with data-theme="e". As a result, the content theme is inheriting the "e" theme from its parent container (see Figure 7–8).

Listing 7–7. *Theme inheritance (ch7/theme-inheritance.html)*

```
<div data-role="page" data-theme="e">
    <div data-role="header">
            <h1>No inheritance</h1>
    </div>

    <div data-role="content">
            Inherits "e"

            <ul data-role="listview" data-inset="true">
                    <li data-role="list-divider">No inheritance</li>
                    <li>No inheritance</li>
                    <li>No inheritance</li>
            </ul>
```

```
                    <form id="test" id="test" action="#" method="post">
                      <p>
                        <label for="text">Inherits "e"</label>
                        <input type="text" name="text" id="text" value=""/>
                      </p>
                      <p>
                        <label for="sound">Inherits "e"</label>
                        <select name="slider" id="sound" data-role="slider">
                            <option value="off">Off</option>
                            <option value="on">On</option>
                        </select>
                      </p>

                        <a href="#" data-role="button">Button (Inherits "e")</a>
                    </form>
            </div>

            <div data-role="footer" data-position="fixed">
                    <h3>No inheritance</h3>
            </div>
    </div>
```

Figure 7–8. *Theme inheritance*

> **NOTE:** Not all components will inherit the theme of their parent container. Refer to the "Inherits Parent Theme" column in Table 7–1 for a listing of components that will not inherit a parent theme.

We can also explicitly set the themes of our individual components. This gives the designer flexibility in regards to styling sites and can help build more rich designs (see Figure 7–9 and its related code in Listing 7–8).

Figure 7–9. *Explicit themes*

Listing 7–8. *Explicit themes (ch7/theme-explicit.html)*

```
<div data-role="page" data-theme="e">
    <div data-role="header" data-theme="b">
      <h1>Theme = "b"</h1>
    </div>

    <div data-role="content" data-theme="d">
      Theme = "d"

      <ul data-role="listview" data-theme="e" data-divider-theme="e">
```

```
            <li data-role="list-divider">Theme = "e"</li>
            <li>Inherits "e" from list</li>
            <li data-theme="b">Theme = "b"</li>
        </ul>

        <form id="test" id="test" action="#" method="post">
        <p>
            <label for="text">Theme "d"</label>
            <input type="text" name="text" id="text" data-theme="d" />
        </p>
        <p>
          <label for="sound">Theme "b"</label>
          <select id="sound" data-role="slider" data-theme="b">
            <option value="off">Off</option>
            <option value="on">On</option>
          </select>
        </p>

        <a href="#" data-role="button" data-theme="a">Button</a>
        </form>
    </div>

    <div data-role="footer" data-position="fixed" data-theme="b">
      <h3>Theme = "b"</h3>
    </div>
</div>
```

Theme Precedence

Themes are applied to components with the following order of precedence:

1. Explicit themes—If you explicitly set the data-theme attribute on any component, that theme will override any inherited or default theme.

2. Inherited themes—Inherited themes will override all default themes. For instance, in Listing 7–7, the content container inherited theme "e" from its page container which overrode its default theme "c". For a list of components that may inherit their theme, refer to the "Inherit's Parent Theme" column in Table 7–1.

3. Default themes—Default themes are applied when no themes are explicitly set or inherited. For a listing of default themes by component, refer to the "Default Theme" column in Table 7–1.

TIP: By default, the minimum height of the content container will only stretch the height of the components inside. This is an issue when the theme of the content is different than the theme of its page container (see Figure 7–10). We can remedy this issue with CSS. For instance, we can set the minimum height of our content container to the height of the screen (see Figure 7–11):

```
.ui-content { min-height:inherit; }
```

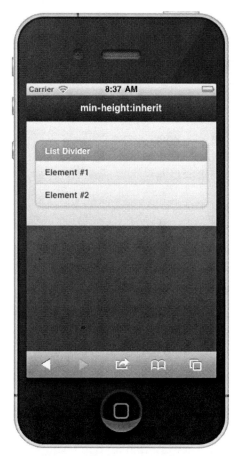

Figure 7-10. *Content height not 100%*

Figure 7-11. *Content height 100% (ch7/min-height.html)*

Custom Themes

The jQuery Mobile theming framework allows designers to quickly customize or re-style their user interface. In this section, we will see how we can manually create our own custom swatches. As reviewed previously, the default jQuery Mobile CSS document is broken into two sections: a theme section and a structure section. For this exercise we are going to create a custom swatch that we can reference for potentially dangerous actions. For instance, a common user experience guideline encourages developers to highlight buttons that control potentially harmful actions in red. In jQuery Mobile, we can create a custom swatch to manage the visual styling (backgrounds, borders, color, font, shadows) of icons and/or buttons that drive our risky actions.

To create a custom swatch manually the following steps are necessary:

1. First, create a separate CSS file for the custom theme (css/theme/custom-theme.css). This keeps the custom additions separate from the main jQuery Mobile CSS and will simplify future upgrades.

> **TIP:** If you plan to style your entire jQuery Mobile application with custom themes it is recommended to use the structure-only CSS file from jQuery Mobile's download site2. This is a lightweight alternative for applications that do not need the default themes and it simplifies the management of the custom themes (see Listing 7-9).

Listing 7-9. *jQuery Mobile's structure file without default themes*

```
<head>
  <meta charset="utf-8">
  <title>Custom Theme</title>
  <meta name="viewport" content="width=device-width, initial-scale=1">
  <link rel=stylesheet href="css/theme/custom-theme.css" />
  <link rel=stylesheet href="css/structure/jquery.mobile.structure.css"/>
  <script type="text/javascript" src="jquery-min.js"></script>
  <script type="text/javascript" src="jquery.mobile-min.js"></script>
</head>
```

2. Find an existing swatch to reference as a baseline. After studying the existing swatches, copy one that will closely resembles the style of your new swatch. This will help minimize the number of modifications you will have to make in order to create your new swatch. For my new swatch, I copied the "e" swatch as my baseline because "e" is an accent swatch and our new swatch for potentially dangerous actions can be segmented into the accent category too.

3. Next, copy the baseline swatch and paste it into the custom-theme.css file. Then, rename the swatch so it is associated to a unique letter (f-z). For example, replace all CSS suffixes with "-e" to "-v" (see Listing 7-10). The new swatch can now be referenced with data-theme="v" for any components that perform dangerous actions.

Listing 7-10. *Custom "v" swatch modeled after swatch "e" (ch7/css/theme/custom-theme1.css)*

```
/* V
---------------------------------------------------------------------*/
.ui-bar-v {
  font-weight: bold;
  border: 1px solid      #999;
  background:            #dedede;
  color:                 #000;
```

---

[2] See http://jquerymobile.com/download.

```
  text-shadow: 0 1px 0px  #fff;
  …
}
.ui-btn-up-v {
  border: 1px solid       #999;
  background:             #e79696;
  color:                  #fff;
  text-shadow: 0 1px 0px  #fff;
  ...
}
```

4. Now the exciting task of updating the visual CSS settings (backgrounds, borders, color, font, and shadows) for our new swatch. For the new "v" swatch, I updated all buttons to have a red gradient background with white text (see Listing 7-11).

Listing 7-11. *Update "v" swatch buttons with red background gradient and white text (ch7/css/theme/custom-theme1.css)*

```
/* V
------------------------------------------------------------------*/
.ui-btn-up-v {
  border: 1px solid       #999;
  background:             #e79696;

  color:                  #fff;

  text-shadow: 0 1px 0px  #fff;
 background-image: -webkit-gradient(
        linear, 0% 0%, 0% 100%, from(#E79696), to(#ce2021),
        color-stop(.4,#E79696)
  );
  background-image: -webkit-linear-gradient(
        0% 56% 90deg,#CE2021, #E79696, #E79696 100%
  );
  background-image:    -moz-linear-gradient(
        0% 56% 90deg,#CE2021, #E79696, #E79696 100%
  );
  …
}
```

5. Next, we need to integrate our new "v" swatch with an actual page for testing. I created two pages to help test the new "v" swatch. On the first page, I wanted to see how the new swatch appeared on an icon-only button. For this test, I created a split button list with the secondary button as our delete icon and styled the secondary button with our new "v" swatch (see Figure 7-12 and its related code in listing 7-12).

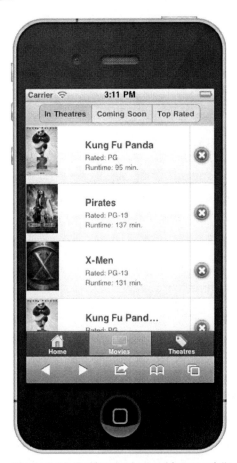

Figure 7-12. *Red icon background for potentially dangerous actions*

Listing 7-12. *Split buttons with "v" swatch (ch7/custom1.html)*

```
<head>
  <link rel=stylesheet href="css/theme/custom-theme1.css" />
  <link rel=stylesheet href="css/structure/jquery.mobile-min.css"/>
  ...
</head>

<ul data-role="listview" data-split-icon="delete" data-split-theme="v">
    <li>
        <a href="#">
            <img src="../images/kungfupanda2.jpg" />
            <h3>Kung Fu Panda</h3>
            <p>Rated: PG</p>
            <p>Runtime: 95 min.</p>
        </a>
    <a href="#delete" data-transition="slidedown">Delete</a>
    </li>
...
```

I also imported our new custom-theme.css file before the main jQuery Mobile CSS file. For our second test, I wanted to apply our new "v" swatch on a delete button. For this test, I created a dialog to confirm the potentially dangerous action and styled the delete button with the our new red gradient styled theme (see Figure 7-13 and its related code in Listing 7-13).

Figure 7-13. *Red delete button for delete action*

Listing 7-13. *Delete button with "v" swatch (ch7/custom1.html)*

```
<div data-role="dialog" id="delete">
    <div data-role="content" data-theme="c">
        <span class="title">Are you sure?</span>

        <a href="#home" data-role="button" data-theme="v">Delete</a>
        <a href="#home" data-role="button" data-rel="back">Cancel</a>
    </div>
</div>
```

Lastly, as you create new swatches, it will be helpful to document your custom styles within a color-coded style guide so all designers and developers within the organization will be familiar with their usage and style.

> **TIP:** CSS gradient generators[3] are tools that can automate the generation of your gradient syntax and help simplify step 4.

ThemeRoller

ThemeRoller[4] is a web-based tool that helps automate the process of generating new CSS-based themes for jQuery Mobile. This is a very helpful tool because it allows you to make color scheme updates in the left pane and preview the results in the right pane within an actual jQuery Mobile layout (see Figure 7-14).

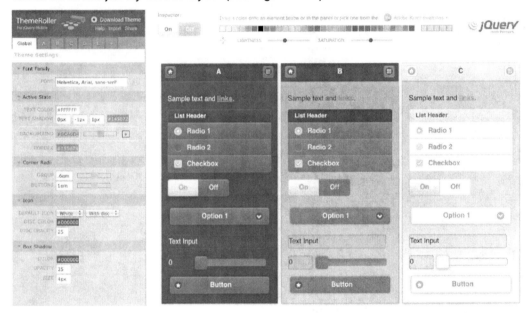

Figure 7-14. *ThemeRoller*

---

[3] See `http://www.westciv.com/tools/gradients/` or `http://gradients.glrzad.com/`.

[4] See `http://jquerymobile.com/themeroller`.

Swatch and Global Settings

You can quickly adjust the CSS attributes that apply globally to all swatches under the "Global" tab that appears in the left pane. Here you can adjust the font family, active state colors, corner radii, icons, and shadows (see Figure 7-15).

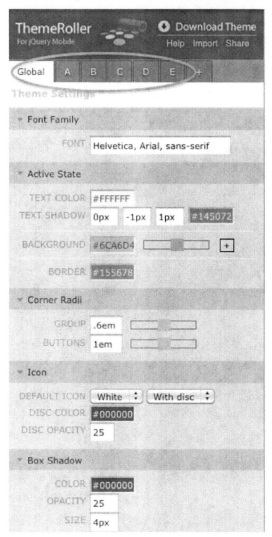

Figure 7-15. *Global Theme Settings*

Next to the "Global" tab are the swatch specific tabs (a-z). Here you may add, edit or delete a swatch from your theme (see Figure 7-15).

Preview Inspector and QuickSwatch Bar

To make it even easier to build custom themes there are two unique tools at the top of the preview panel: Preview Inspector and the QuickSwatch Bar.

The Preview Inspector is a toggle that can be either "On" or "Off" (see Figure 7-16). With the toggle "On", clicking on an element in the preview pane will automatically show its editable attributes in the left pane. This will be a valuable timesaver when you need to quickly edit styles.

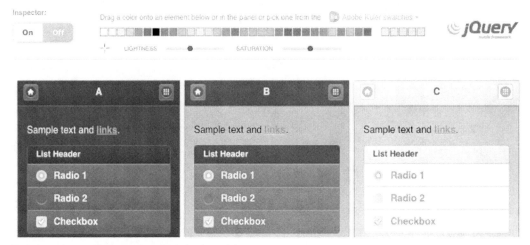

Figure 7-16. *Preview inspector and QuickSwatch bar*

The QuickSwatch Bar is a spectrum of colors that appears to the right of the Inspector (see Figure 7-16). This is a powerful tool that allows you to drag and drop any color onto an element in the preview page or onto a color attribute in the left pane. Below the QuickSwatch bar are two sliders to adjust the lightness and saturation of your color pallet. Additionally, the most recent selected colors will be shown to the right of the color spectrum for quick reuse.

Adobe Kuler Integration

It can be challenging when you need to create a color pallet from scratch. To help simplify this process, ThemeRoller has Adobe's Kuler[5] integration built-in (see Figure 7-17).

Figure 7-17. *Adobe's Kuler App*

Kuler is a site that allows people to create, share, and rate color pallets. To see the color pallets that are available in Kuler click the "Adobe Kuler" link that appears above the QuickSwatch Bar. When the Kuler app opens, a search filter appears in the left pane that allows you to filter by latest, popular, rating, or custom search. When you find a color of interest, simply drag and drop the color onto an element in your preview pane.

Getting Started

For comparison purposes, I am going to create a red accent swatch in ThemeRoller to see how this experience compares to the manual swatch we created in the previous section. In this exercise, I am going to override jQuery Mobile's default "e" swatch with the new red accent swatch. In ThemeRoller, to update an existing theme the following steps are necessary:

1. In ThemeRoller, import an existing theme by clicking the "Import" link in the upper left corner (see Figure 7-18). For this exercise I am going to import and modify jQuery Mobile's default theme.

Figure 7-18. *Import Existing Theme*

2. After the theme is imported, identify the swatch to modify. For this step, I am going to modify the default "e" swatch.

3. Next, find an appropriate base color for our red accent swatch. We can find an appropriate red color in either the QuickSwatch Bar or Kuler integration tool.

---

[5] See http://kuler.adobe.com.

4. After finding an appropriate base color we can now update the elements in our preview pane with the chosen color. For example, I am going to style the header and all elements with a deep red accent color.

5. In the preview pane, make any necessary adjustments. For instance, you may want to adjust the colors slightly or add subtle effects with background gradients. As expected, ThemeRoller makes the editing and preview process much more efficient compared to the manual approach!

6. After you are comfortable with the layout of the new theme, you can download the CSS of the theme by clicking on the "Download Theme" link in the upper left corner of ThemeRoller (see Figure 7-19).

Figure 7-19. *Download Theme*

7. We can now begin referencing the new theme in our application (see Listing 7-14 and its related screenshot in Figure 7-20). Again, to help simplify the management of the custom themes it is recommended to load the structure file and your custom themes separately.

Listing 7-14. *ThemeRoller custom theme import (ch7/custom2.html)*

```
<head>
  <meta charset="utf-8">
  <title>Custom Theme</title>
  <meta name="viewport" content="width=device-width, initial-scale=1">
  <link rel=stylesheet href="css/theme/custom-theme2.css" />
  <link rel=stylesheet href="css/structure/jquery.mobile.structure.css"/>
  <script type="text/javascript" src="jquery-min.js"></script>
  <script type="text/javascript" src="jquery.mobile-min.js"></script>
</head>

<div data-role="dialog" id="delete">
    <div data-role="content" data-theme="c">
        <span class="title">Are you sure?</span>

        <a href="#home" data-role="button" data-theme="e">Delete</a>
        <a href="#home" data-role="button" data-rel="back">Cancel</a>
    </div>
</div>
```

Figure 7-20. ThemeRoller's red delete button

Summary

The jQuery Mobile theming framework is an object-oriented CSS3 framework that is lightweight, customizable, and renders unified designs across all browsers. In this chapter, we discussed the basics of the theming framework and reviewed the five swatches that are included in jQuery Mobile.

We also reviewed the three ways themes can be assigned to components. While all components can have their theme explicitly set with the data-theme attribute, most components also have default themes and can also inherit their theme from a parent container. We saw examples of each and reviewed the precedence in which themes are applied to components.

Lastly, we saw how to create our own custom swatches. Whether you need to create a richer design or you need to create a design that closely matches your corporate branding, the theming framework is flexible enough for all requirements. We reviewed the two options available for creating custom swatches and saw step-by-step examples of each. We saw that the manual approach gives us complete control of our layout, and that jQuery Mobile's new ThemeRoller provides a more efficient, intuitive working environment.

In our next chapter, we are going to take an in-depth look at the jQuery Mobile API. We will learn how to configure jQuery Mobile and we will also review the API's core methods, events, properties, and data attributes.

jQuery Mobile API

All well-written frameworks allow developers to extend and override default configuration settings. Additionally, they provide convenience methods to help simplify your code. jQuery Mobile includes a fairly extensive API that exposes each of these convenient features. First, we will look at how to configure jQuery Mobile. We will review each configurable feature within jQuery Mobile, highlight its default setting, and show how the API allows you to configure each option. Then, we will explore the most popular methods, page events, and properties that jQuery Mobile exposes. These API features are useful when you need to programmatically update your Mobile Web applications. Lastly, we will review a sorted table listing of all jQuery Mobile data attributes. For each attribute, we will include a brief description, example, and figure of its enhanced component.

Configuring jQuery Mobile

When jQuery Mobile initializes, it triggers a `mobileinit` event on the document object. You can bind to the `mobileinit` event and apply overrides to jQuery Mobile's (`$.mobile`) default configuration settings. In addition, you can extend jQuery Mobile with additional behavior and properties. For instance, there are two ways to configure jQuery Mobile as shown in the examples below. You can either override the properties via jQuery's extend method or individually.

Examples:

```
// Configure properties via jQuery's extend method
$( document ).bind( "mobileinit", function(){
  $.extend( $.mobile, {
    // Override loading message
    loadingMessage: "Loading...",

    // Override default transition from "slide" to "pop"
    defaultTransition: "pop"
  });
});
```

```
// Configure properties individually
$( document ).bind( "mobileinit", function(){
  $.mobile.loadingMessage = "Initializing";
  $.mobile.defaultTransition = "slideup";
});
```

Custom Script Placement

Since `mobileinit` gets triggered immediately upon execution of jQuery Mobile, you need to place your custom script before the jQuery Mobile JavaScript file.

Example:

```
<head>
    <script type="text/javascript" src="jquery-min.js"></script>
    <script type="text/javascript" src="custom-scripts-here.js"></script>
    <script type="text/javascript" src="jquery.mobile-min.js"></script>
</head>
```

Configurable jQuery Mobile Options

The following are configurable `$.mobile` options you may override within your custom JavaScript.

- **activeBtnClass**(string, default: "ui-btnactive")

  The CSS class used to identify and style the "active" button. This CSS attribute is commonly used to style and identify the active button in a tab bar.

- **activePageClass**(string, default: "ui-page-active")

  The CSS class assigned to the page or dialog that is currently visible and active. For instance, when multiple pages are loaded in the DOM the active page will have this CSS attribute applied.

- **ajaxEnabled**(boolean, default: true)

  Dynamically load pages via Ajax when possible. Ajax loading is on by default for all pages except external URL's, or links that are marked with a `rel="external"` or `target="_blank"` attribute. If Ajax is disabled, page links will be loaded with regular HTTP requests without CSS transitions.

- **allowCrossDomainPages**(boolean, default: false)

  When developing with PhoneGap, it is recommended to set this configuration option to true. This allows jQuery Mobile to manage the page loading logic of cross-domain requests in PhoneGap.

▓ **autoInitializePage**(boolean, default: true)

For advanced developers that want total control of the initialization sequence of a page, you may set this config option to `false`, which disables auto-initialization of all page components. This allows developers to manually enhance each control on demand.

▓ **defaultDialogTransition**(string, default: "pop")

The default transition to use when transitioning to a dialog. You may set the transition to "none" for no transition.

▓ **defaultPageTransition**(string, default: "slide")

The default transition to use when transitioning to a page. You may set the transition to "none" for no transition.

▓ **gradeA**(function that returns a boolean, default: browser must support media queries or support IE 7+);

jQuery Mobile will call this method to determine if the framework will apply dynamic CSS page enhancements. By default, the method will apply enhancements for all browsers that support media queries; however, jQuery Mobile will only enhance pages for grade-A browsers. IE 7 and above are included as grade-A browsers and their displays will be enhanced too For instance, this is the current function for `$.mobile.gradeA`:

```
$.mobile.gradeA: function(){
  return $.support.mediaquery ||
         $.mobile.browser.ie && $.mobile.browser.ie >= 7;
}
```

▓ **hashListeningEnabled**(boolean, default: true)

Automatically load and show pages based on `location.hash`. jQuery Mobile listens for `location.hash` changes to load internal pages within the DOM. You can disable this and handle the hash changes manually or disable this option to access an anchor's bookmark as a deep link.

▓ **loadingMessage**(string, default: "loading")

Sets the loading message to appear during Ajax-based requests. Additionally, you can assign a `false` (boolean) to disable the message. Also, if you want to update the loading message at runtime on a per-page basis you can update it within your page.

Example:

```
// Update loading message
$.mobile.loadingMessage = "My custom message!";

// Show loading message
$.mobile.showPageLoadingMsg();
```

▓ **minScrollBack**(string, default: 250)

Sets the minimum scroll distance that will be remembered when returning to a page. When returning to a page, the framework will automatically scroll to the position or link that launched the transition when the scroll position of that link is beyond the minScrollBack setting. By default, the scroll threshold is 250 pixels. If you want to eliminate the minimum setting so the framework always scrolls regardless of the scroll position, set this value to "0". If you want to disable this feature, set the value to "infinity".

▓ **nonHistorySelectors**(string, default: "dialog")

You can specify which page components to exclude from the browser's history stack. By default, any link with data-rel="dialog" or any page with data-role="dialog" will not appear in history. Furthermore, these non-history selector components will not have their URL updated when navigating to their page and as a result they will not be bookmarkable.

▓ **ns**(string, default: "")

The namespace for custom data-* attributes within jQuery Mobile. Data attributes are a new feature within HTML5. For instance, "data-role" is the default namespace for the role attribute. If you wanted to override the default namespace globally you would override the $.mobile.ns option.

Example:

```
// Set a custom namespace
$.mobile.ns = "jqm-";
```

As a result, all of your jQuery Mobile data-* attributes will require the prefix "data-jqm-". For instance, the "data-role" attribute now becomes "data-jqm-role".

> **IMPORTANT:** If you update the default namespace you will need to update one CSS
> selector found within the jQuery Mobile CSS file:
>
> ```
> // Original CSS for default namespace:
> .ui-mobile [data-role=page],
> .ui-mobile [data-role=dialog],
> .ui-page {..}
> ```
>
> ```
> // Updated CSS for the new namespace "jqm-":
> .ui-mobile [data-jqm-role=page],
> .ui-mobile [data-jqm-role=dialog],
> .ui-page {..}
> ```

Why override the default namespace?

First off, if you are designing a JavaScript framework that includes HTML5
data-* attributes, the W3C recommends that you include a hook to allow the
developers to customize namespaces to avoid collisions with third party
frameworks. And anytime you encounter a namespace collision with another
third-party framework you will need to change your default namespace.

■ **page.prototype.options.addBackBtn**(boolean, default: false)"

If you want the back button to appear across your application, set this
option to true. The back button within jQuery Mobile is a smart
widget. It will only appear when there is a page in the history stack to
go back to.

Example:

```
$.mobile.page.prototype.options.addBackBtn = true;
```

■ **page.prototype.options.keepNative**(string, default:

:jqmData(role='none'),:jqmData(role='nojs')"

If you want to prevent auto-initialization without adding data-
role="none" to your markup, you can customize the keepNative
selector that is used for preventing auto-initialization. For instance, to
prevent the framework from initializing all select and input elements we
can update this selector.

Example:

```
$.mobile.page.prototype.options.keepNative = "select, input";
```

■ **pageLoadErrorMessage**(string, default: "Error Loading Page")

The error response message that appears when an Ajax page request fails to
load.

▨ **subPageUrlKey**(string, default: "ui-page")

The URL parameter used for referencing widget-generated sub-pages. An example of a sub-page URL would appear as "*nested-list.html&ui-page=Movies-3*". A nested list view is a particular widget which segments each list into individual sub-pages. For example, the URL shown previously has a "Movies" sub-list that jQuery Mobile transformed into its own subpage to accommodate a deep link reference. If you need to rename this URL parameter, you can change it with `$.mobile.subPageUrlKey`.

▨ **touchOverflowEnabled**(boolean, default: false)

In order to achieve true fixed toolbars with native momentum scrolling, a browser needs to either support position:fixed or overflow:auto. Fortunately, new releases of WebKit (iOS5) are beginning to support this behavior. It is very likely that this option will become enabled by default. Until this occurs, we can enable this behavior by setting this configuration option to true.

Methods

jQuery Mobile exposes a suite of methods that are helpful when you need to programmatically update your Mobile Web application.

▨ **$.mobile.changePage()**

The changePage function handles all the details of transitioning from one page to another.

Usage

`$.mobile.changePage( toPage, [options] )`

Arguments

▨ toPage (sting or jQuery collection). The page to transition to.

▨ toPage (string). A file URL ("contact.html") or internal element's ID ("#contact").

▨ toPage (object). A jQuery collection object containing a page element as its first argument: $("#contactPage")

options (object). A set of key/value pairs that configure the changePage request. All settings are optional.

▨ allowSamePageTransition (boolean, default: false). The changePage method will ignore requests that transition to the same page. Set this option to `true` to allow same page transitions.

- changeHash (boolean, default: true). Update the hash to the toPage's URL when the page change is complete.

- data (string or object, default: undefined). The data to send to an Ajax page request.

- dataUrl (string, default: toPage URL). Sets the URL to show in the browser's location field.

- fromHashChange (boolean, default: false). To indicate if the changePage came from a hashchange event.

- fromPage (string, default: $.mobile.activePage). Specifies the from page.

- pageContainer (jQuery collection, default: $.mobile.pageContainer). Specifies the element that should contain the page after it is loaded.

- reloadPage (boolean, default: false). Force a reload of the page even if it is already in the DOM of the page container.

- reverse (boolean, default: false). To indicate if the transition should go forward or reverse. The default transition is forward.

- role (string, default: "page"). The data-role value to be used when displaying the page. For dialogs use "dialog".

- showLoadMsg (boolean, default: true). Display the loading message when a page is requested.

- transition (string, default: $.mobile.defaultTransition).The transition to apply for the change page. The default transition is slide.

- type (string, default: "get"). Specifies the method ("get" or "post") to use when making a page request.

Example #1:

```
//Transition to the "contact.html" page.
$.mobile.changePage( "contact.html" );

<!-- Markup equivalent when link clicked -->
<a href="contact.html">Contact Us</a>
```

Example #2:

```
// Transition to the internal "#contact" page with a reverse "pop" transition.
$.mobile.changePage( "#contact", { transition: "pop", reverse: true } );

<!-- Markup equivalent when link clicked -->
<a href="contact.html" data-transition="pop" data-
direction="reverse">Contact</a>
```

Example #3:

```
/* Dynamically create a new page and open it */

// Create page markup
var newPage = $("<div data-role=page data-url=hi><div data-role=header>
    <h1>Hi</h1></div><div data-role=content>Hello Again!</div></div>");

// Add page to page container
newPage.appendTo( $.mobile.pageContainer );

// Enhance and open new page
$.mobile.changePage( newPage );
```

$.mobile.hidePageLoadingMsg()

Remove or hide the page loading message
($.mobile.loadingMessage). The default loading message is "loading"
and this is configurable as well. To show the loading message, see
$.mobile.showPageLoadingMsg().

Example:

```
// Remove the loading message
$.mobile.hidePageLoadingMsg();
```

$.mobile.loadPage()

The loadPage function loads a page into the DOM of the current page and
enhances it. This method is also exposed as a data attribute and can be
attached to links or buttons (see "data-prefetch", in Data Attributes Section).

Usage

```
$.mobile.loadPage( url, [options] )
```

Arguments

url (sting). The page to load.

- url (string). A file URL ("contact.html").

options (object). A set of key/value pairs that configure the changePage request.
All settings are optional.

- data (string or object, default: undefined). The data to send to an
  Ajax page request.

- loadMsgDelay (number (in ms), default: 50). Add a manual delay
  before the loading message is shown. This delay allows the
  framework to load a cached page without a loading message.

- PageContainer (jQuery Collection, default:
  $.mobile.pageContainer).The element that should contain the
  page after it is loaded.

- reloadPage (boolean, default: false). Force a reload of the page even if it is already in the DOM of the page container.

- role (string, default: @data-role attribute). The data-role to load the page with. The default is the @data-role attribute defined on the element.

- showLoadMsg (boolean, default: true). Display the loading message when a page is requested.

- type (string, default: "get"). Specifies the method ("get" or "post") to use when making a page request.

Examples:

```
// Dynamically load a page and transition to it.
$.mobile.loadPage("page1.html" );

$.mobile.changePage("#page1" ); // data-url value
```

- **$.mobile.showPageLoadingMsg()**

Show the page loading message ($.mobile.loadingMessage).

Example:

```
// Show the page loading message
$.mobile.showPageLoadingMsg();
```

- **$.mobile.silentScroll(number)**

Scrolls the page vertically. Within the framework, silentScroll is called whenever a page gets restored. For example, when you click on the back button the silentScroll method gets triggered before the prior page is shown and will restore the prior page's scroll position. Focus will be on the component that triggered the initial transition. The scrollstart and scrollstop events will not get triggered during a silentScroll.

Example:

```
// Hide the iOS address bar
$.mobile.silentScroll(0);

// Scroll down 400 pixels
$.mobile.silentScroll(400);
```

■ **$.jqmData()**

This is the mobile version of the jQuery .data() method.[1] This method provides all functionality found within $.data() plus it ensures all data is set and retrieved using jQuery Mobile's data namespace ($.mobile.ns).

Examples:

```
// Find all pages (data-role="page") in the DOM via a selector.
var $pages = $( ":jqmData(role='page')" );

// Find the theme (data-theme) for the first page
var firstPage = $pages.first();
var theme = $.jqmData( firstPage, "theme" );
```

■ **$.jqmHasData()**

This is the mobile version of the jQuery .hasData() method.[2] This method provides all functionality found within $.hasData() plus it ensures all data is retrieved using jQuery Mobile's data namespace ($.mobile.ns).

Examples:

```
// Does a theme exist for the first page
var hasTheme = $.jqmHasData( firstPage, "theme" );
```

■ **$.jqmRemoveData()**

This is the mobile version of the jQuery .removeData() method.[3] This method provides all functionality found within $.removeData() plus it ensures all data is removed using jQuery Mobile's data namespace ($.mobile.ns).

Examples:

```
// Set data on the first page
$.jqmData(firstPage, "testData", "testValue");

// Remove the data from the first page
$.jqmRemoveData( firstPage, "testData" );
```

---

[1] http://api.jquery.com/jQuery.data/

[2] http://api.jquery.com/jQuery.hasData/

[3] http://api.jquery.com/jQuery.removeData/

Events

jQuery Mobile also exposes several events that are helpful when you need to programmatically apply pre or post processing during page changes within your Mobile Web application. In this section we will review the complete list of jQuery Mobile page events that you may bind to in your own code. For an introduction of the jQuery Mobile events lets begin with a diagram (Figure 8-1). This diagram shows the main page events that occur within jQuery Mobile and helps depict the sequence of each event in the page change lifecycle.

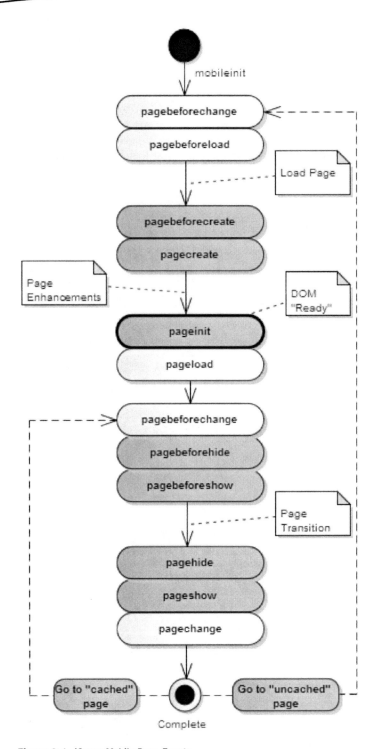

Figure 8-1. *jQuery Mobile Page Events*

Now that we have seen the sequence in which page events are triggered within the page change lifecycle, let's look at the details of each specific event.

Events Overview

▓ **mobileinit**

When jQuery Mobile initializes, it triggers a `mobileinit` event on the document object. You can bind to the `mobileinit` event and apply overrides to jQuery Mobile's default configuration settings. Refer to Section, "Configuring jQuery Mobile", for examples of binding to the `mobileinit` event.

Page Change Events

Page change events are automatically triggered on the document when you navigate to another page. Internally, these events are trigger when the `$.mobile.changePage` method is called. During this process, two events will be fired. The first event triggered is pagebeforechange. The next event to fire is dependent upon the status of the page change. When the page change is successful, the `pagechange` event will be triggered and if the page change fails, the pagechangefailed event fires.

▓ **pagebeforechange**

This is the first event that gets triggered during a page change. Callbacks for this event are passed two arguments. The first is the event and the second argument is a data object. You may cancel the page change by calling preventDefault on the event. Additionally, you can override the page change by inspecting and updating the data object. The data object, passed as the second argument, contains the following properties:

▓ toPage (string). A file URL or a jQuery Collection object. This is the same argument that was passed to $.mobile.changePage().

▓ options (object). These are the same options that were passed to $.mobile.changePage.

Example:

```
$( document ).bind( "pagebeforechange", function( e, data ) {
    console.log("Change page starting...");

    // Get the page
    var toPage = data.toPage;

    // Get the page options
    var options = data.options;

    // Inspect toPage or override options (redirect)…

    // Prevent a page change
    e.preventDefault();
});
```

▓ **pagechange**

This is the last event to trigger after a successful page change. Callbacks for this event are passed two arguments. The first is the event and the second argument is a data object. The data object, passed as the second argument, contains the following properties:

▓ toPage (string). A file URL or a jQuery Collection object. This is the same argument that was passed to `$.mobile.changePage()`.

▓ options (object). These are the same options that were passed to `$.mobile.changePage`.

Example:

```
$( document ).bind( "pagechange", function( e, data ){
    console.log("Change page successfully completed...");
    var toPage = data.toPage;
    var options = data.options;
});
```

▓ **pagechangefailed**

This event is triggered if the page change fails. The callbacks for this event are passed two arguments. The first is the event and the second argument is a data object. The data object, passed as the second argument, contains the following properties:

▓ toPage (string). A file URL or a jQuery Collection object. This is the same argument that was passed to `$.mobile.changePage()`.

▓ options (object). These are the same options that were passed to `$.mobile.changePage`.

Example:

```
$( document ).bind( "pagechangefailed", function( e, data ){
    console.log("Page change failed...");
});
```

Page Load Events

Page load events get triggered on the document when the framework loads a page into the DOM. Programmatically, this event is trigger when `$.mobile.loadPage` is called. During this process, `loadPage()` will fire off two events. The first is pagebeforeload and the second event is either a success (pageload) or failure (pageloadfailed) event.

▓ **pagebeforeload**

This is the first event to trigger during a page load. Callbacks for this event are passed two arguments. The first is the event and the second argument is a data object. You can manually handle the loading logic if you choose. To accomplish this, you must call `preventDefault()` on the event and call either `resolve()` or `reject()` on the deferred object reference contained in the data object. The data object, passed as the second argument, contains the following properties:

▓ `url` (string). The relative URL that was sent to `$.mobile.loadPage()`.

▓ `absUrl` (string). An absolute reference of the URL.

▓ `dataUrl` (string). The version of the URL actually stored in the data-url attribute of the page. This URL is shown in the browser's location field.

▓ `deferred` (object). Callbacks that call `preventDefault()` to manually handle the page loading must call `resolve()` or `reject()` on this object so the `changePage()` request can resume processing.

▓ `options` (object). This is the same options argument that was passed to `$.mobile.loadPage()`.

Example:

```
$( document ).bind( "pagebeforeload", function( e, data ){
    console.log("Page load starting…");

    // Let the framework know we're manually loading the page
    e.preventDefault();

    // Manually load the document and insert it into the DOM
    var response = manuallyLoadPage();

    if (response.status = "success") {
        // Call resolve passing in the url, options, and jQuery
        // collection object containing the DOM element for the page
        data.deferred.resolve( data.absUrl, data.options,
            response.page);
    } else {
        // The load failed, call reject
        data.deferred.reject( data.absUrl, data.options );
    }
});
```

▓ **pageload**

This event is triggered after the page is successfully loaded into the DOM. Callbacks for this event are passed two arguments. The first is the event and the second is a data object. The data object, passed as the second argument, contains the following properties:

- url (string). The relative URL that was sent to
  `$.mobile.loadPage`.

- absUrl (string). An absolute reference of the URL.

- dataUrl (string). The version of the URL actually stored in the
  data-url attribute of the page. This URL is shown in the browser's
  location field.

- options (object). This is the same options argument that was
  passed to `$.mobile.loadPage()`.

Example:

```
$( document ).bind( "pageload", function( e, data ){
  console.log("Page successfully loaded into DOM...");
});
```

- **pageloadfailed**

- This event is triggered if the page load fails. During this process, the
  framework will display a page failed message and call `reject()` on the
  deferred object. Callbacks can prevent this default behavior from
  executing by calling `preventDefault()` on the event.

Example:

```
$( document ).bind( "pageloadfailed", function( e, data ){
  console.log("Page load failed...");
});
```

Page Initialization Events

Page initialization events get triggered on the page before and after the jQuery Mobile
enhances the page. You may bind to these events to pre-parse the markup before the
framework enhances the page or afterwards to setup DOM ready event handlers. These
events are only triggered once during the lifecycle of a page.

- **pagebeforecreate**

  Triggered on the page being initialized during a page change. This
  event occurs after the page container has been inserted into the DOM,
  but before the page has been enhanced. This is the preferred location
  to pre-parse the markup before the framework enhances the page.
  For instance, in this event you may dynamically create and append
  new page widgets or modify existing data-attributes.

  Example:

```
$( "#to-page-id" ).live( "pagebeforecreate", function(){
console.log( "Pre-parse the markup before the framework enhances the widgets" );
});
```

- **pagecreate**

  Triggered on the page that is being initialized during a page change. This is the event that gets triggered by the framework to initialize all page plugins. If you create custom page plugins this is the preferred location to initialize them.

  Example:

  ```
  $( "#to-page-id" ).live( "pagecreate", function(){
      console.log("Page plugins are being initialized...");

      // Initialize custom plugins
      ( ":jqmData(role='my-plugin')" ).myPlugin();
  });
  ```

- **pageinit**

  Triggered on the page that is being initialized after enhancements are complete. The page is now in a DOM ready state.

  Example:

  ```
  $( "#to-page-id" ).live( "pageinit", function(){
      console.log("The page has been enhanced...");
      // Attach event handlers or run other jQuery code...
  });
  ```

Page Transition Events

Page transition events get triggered on the "from" and "to" pages during a page transition. You may bind to these events to observe when pages are being shown or removed from view.

- **pagebeforehide**

  Triggered on the "from" page as the transition begins. This event occurs before the pagebeforeshow event. This event will only fire if the page change request has an associated "from" page. Callbacks for this event are passed two arguments. The first is the event and the second argument is a data object. The data object, passed as the second argument, contains the following properties:

    - nextPage (object). A jQuery collection object containing the page element we are transitioning to.

  Example:

  ```
  $( "#from-page-id" ).live( "pagebeforehide", function( e, data ){
    console.log( "The page transition is just starting..." );
  });
  ```

▪ **pagebeforeshow**

Triggered on the "to" page after the page has been enhanced and just before the page transition begins. Callbacks for this event are passed two arguments. The first is the event and the second argument is a data object. The data object, passed as the second argument, contains the following properties:

▪ prevPage (object). A jQuery collection object containing the page element we are transitioning from.

Example:

```
$( "#to-page-id" ).live( "pagebeforeshow", function( e, data ){
  console.log( "The page transition is just starting..." );
});
```

▪ **pagehide**

Triggered on the "from" page after the transition is complete and before the pageshow event. This event will only fire if the page change request has an associated "from" page. Callbacks for this event are passed two arguments. The first is the event and the second argument is a data object. The data object, passed as the second argument, contains the following properties:

▪ nextPage (object). A jQuery collection object containing the page element we are transitioning to.

Example:

```
$( "#from-page-id" ).live( "pagehide", function( e, data ){
  console.log( "The page transition is complete!" );
});
```

▪ **pageshow**

Triggered on the "to" page after the transition is complete and after the "from" page is hidden. Callbacks for this event are passed two arguments. The first is the event and the second argument is a data object. The data object, passed as the second argument, contains the following properties:

▪ prevPage (object). A jQuery collection object containing the page element we are transitioning from.

Example:

```
$( "#to-page-id" ).live( "pageshow", function( e, data ){
  console.log( "The page transition is complete!" );
});
```

> **TIP:** The jQuery Mobile team has created a helpful bookmarklet that allows you to view the page event history from your browser console (see Figure 8-2). As you navigate your jQuery Mobile application you will be able to view the history of events by page, URL, and timestamp. To install, go to the jQuery Mobile event logger page[4] and follow their instructions for installing the bookmark.

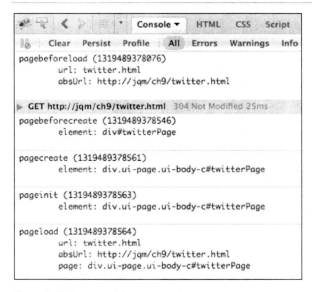

Figure 8-2. *Page event logger console*

Trigger Events

Triggering jQuery Mobile page events can be helpful when building dynamic pages. For instance, if you add several new components to a page you can call the `create` event to enhance all new widgets on the page at once.

- **trigger("create")**

    We can trigger this event to automatically enhance all new elements on a page. This event is triggered on the page container.

    Example:

    ```
    // Add two new buttons to the page
    $( '<button id="b2">Button2</button>' ).insertAfter( "#b1" );
    $( '<button id="b3">Button3</button>' ).insertAfter( "#b2" );

    // Enhance the new buttons on the page
    $.mobile.pageContainer.trigger( "create" );
    ```

---

[4] See http://jquerymobile.com/test/tools/log-page-events.html.

Properties

jQuery Mobile also exposes a set of properties that are publicly available so you do not have to write your own jQuery selectors to access common components.

▓ **$.mobile.activePage**

Gets the page or dialog element that is the currently active or visible. The active page is assigned the CSS class as specified by $.mobile.activePageClass.

▓ **$.mobile.firstPage**

This is the first page defined within the page container ($.mobile.pageContainer). For instance, the $.mobile.firstPage will be shown when no location.hash value exists or when $.mobile.hashListeningEnabled has been disabled. For example, in a multi-page document the $.mobile.firstPage is initially shown by default.

▓ **$.mobile.pageContainer**

The HTML container where all pages live. Within jQuery Mobile, the body element is the container that contains all pages. All Ajax-loaded pages and all internal pages of a multi-page document will exist within the page container.

Data Attributes

jQuery Mobile's data attributes provide the ability to enhance and configure your mobile application with simple HTML markup. The complete list of all data attributes, in alphabetical order with descriptions and examples are listed below (see Table 8-1).

Table 8–1. *jQuery Mobile Data Attribute Reference*

Attribute	Description	Example
data-ajax	This attributed can be attached to links, buttons, or forms. When set to false, it will force a page reload (bypassing Ajax and transitions). For example, this attribute is required on any link that opens a multi-page document from a page that was opened via Ajax. Ajax navigation is enabled by default.	`<a href="multi-page.html" data-role="button" `**`data-ajax="false"`**`>multi-page</a>`
data-add-back-btn	This attributed is attached to a page container. When set to true, a back button will automatically appear in the page header. A page must exist in browser history for the back button to appear. The back button is disabled by default.	◄ Back Contact Us `<div data-role="page" `**`data-add-back-btn="true"`**`>`

Attribute	Description	Example
data-back-btn-text	This attribute is attached to the page container. The default back button text is "Back". You can override this text by updating the value of this attribute.	◄ Previous Contact Us <div data-role="page" data-add-back-btn="true" **data-back-btn-text="Previous"**>
data-collapsed	You can configure a collapsible container to be collapsed (data-collapsed="true") or expanded (data-collapsed="false") by adding this attribute. A collapsible section will be shown expanded by default (see Listing 6-10).	⊕ Wireless ⊖ Wireless 🔲 Notifications ❯ ✈ Location Services ❯ <div data-role="collapsible" data-collapsed="true"> <h3>Wireless</h3> </div>
data-corners	This attribute can be attached to links or buttons. When false, the framework will remove rounded corners from the button. Buttons will have rounded corners by default. For instance, the "disagree" button shown on the right has its rounded corners removed.	*Disagree* *Agree* <a href="" data-role="button" **data-corners="false"**>Disagree</a>
data-count-theme	This attribute sets an alternate theme for your badge or count bubble (see Listing 5-9).	Comments 👤 Thanks for the review. I plan to check it out this weekend. 1 day <ul data-role="listview" **data-count-theme="e"**>
data-direction	This attributed is attached to links, buttons, or forms. When set to reverse it will apply a reverse transition. For example, a forward "slide" transition will slide left. A reverse "slide" transition will slide right. A reverse transition is applied by default when transitioning back in history.	<a href="" data-icon="home" data-iconpos="notext" **data-direction="reverse"** class="ui-btn-right">Home</a>

Attribute	Description	Example
data-divider-theme	Sets the theme of the list divider (see Listing 5-3).	 <li data-role="list-divider" **data-divider-theme="a"**>
data-dom-cache	This attribute allows you to cache pages within the DOM. By default, this attribute is set to false and as a result, the framework will only keep the "from" and "to" pages in the DOM actively cached. It is recommended to leave this value unchanged so the DOM remains lightweight.	<div data-role="page" **data-dom-cache="true"**>
data-filter	This attribute is attached to lists and adds a search bar above the list of results when the value is set to true (see Listing 5-10).	 <ul data-role="listview" **data-filter="true"**>
data-filter-placeholder	Sets the placeholder (hint) text for the search filter. The default placeholder text is "Filter items..." (see Listing 5-10).	 <ul data-role="listview" data-filter="true" **data-filter-placeholder="Search..."**>
data-filter-theme	Sets the theme for the search filter.	 <ul data-role="listview" data-filter="true" **data-filter-theme="e"**>
data-fullscreen	This attribute is attached to a page container. The content section will appear in full-screen mode when set to true. Typically, you will want this behavior for viewing photos and videos (see Listing 3-1).	<div data-role="page" **data-fullscreen="true"**>

Attribute	Description	Example
data-icon	This attribute is attached to links and buttons. For example, setting the value to home shows the home icon from the jQuery Mobile icon pallet. For the complete listing of values refer to Table 4-1.	 <a href="../../" data-icon="home" data-iconpos="notext" data-direction="reverse" class="ui-btn-right jqm-home">Home</a>
data-iconpos	This attribute can be attached to links or buttons. This attribute will position the icon. It can be set to "top", "bottom", "left", "right" or "notext" (see Listing 4-6). "notext", will remove the icon's default text for an icon-only button. By default, icons are left aligned.	 <a href="" *data-icon="home* **data-iconpos="notext"** data-direction="reverse" class="ui-btn-right">Home</a>
data-iconshadow	This attribute can be used in conjunction with the data-icon attribute. When false, the framework will remove the drop shadow from the button's icon. Drop shadows will show by default. For example, the plus icon shown on the right has its shadow removed.	 <a href="#" data-role="button" **data-iconshadow="false"** data-icon="plus" data-iconpos="notext">Plus</a>
data-id	This attribute is attached to the footer and is commonly used with tab bars. Add this attribute to the footer of the active and target page to make the footer stay in place between transitions. The value of this attribute must be the same across pages for the footer to stay in place, and the header and footer toolbars must be set to data-position="fixed". Otherwise they will not be in view during the transition.	<div data-role="footer" **data-id="myFooter"** data-position="fixed">

Attribute	Description	Example
data-inline	This attribute is attached to links or buttons. By default, all buttons in the body content are styled as block-level elements and fill the width of the screen. If you want a more compact button that is only as wide as its text and icons set this value to true. An inline block is placed inline (i.e., on the same line as adjacent content), but it behaves as a block. Adding data-inline to a button will position them side-by-side (see Listing 4-1).	 `<a href="#agree" data-role="button" `**`data-inline="true"`**`>`
data-inset= "true"	This attribute is attached to lists. When set to true it will style the list items so they do not appear edge-to-edge and instead they will appear with rounded corners. Insets help to visually segregate different groups of lists (see Listing 5-2).	`<ul data-role="listview" `**`data-inset="true"`**`>`
data-native-menu	Select menu's will launch the native select picker for the OS by default. To render the select menu in a custom HTML/CSS view, set this value to false.(see Listing 4-17).	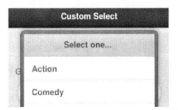 data-native-menu="true" data-native-menu="false" `<select name="genre" `**`data-native-menu="false"`**`>`

Attribute	Description	Example
data-placeholder	A placeholder can be used to display hint text for an unselected select menu and it requires users to make a selection (see "Placeholder Options", Chapter 4).	Ticket Delivery: Select one... `<option value="" data-placeholder="true">Select one...</option>`

data-position	This attribute is attached to headers or footers. When set to `fixed` it will position the header and footer at the top and bottom of the page.	`<div data-role="header" data-position="fixed">`
		
		`<div data-role="footer" data-position="fixed">`

data-prefetch	When this attribute is added to a link or button the framework will lazy load the page into the DOM in the background. It is recommended to build individual pages (single-page template) and use the data-prefetch attribute to preload secondary pages that will be commonly accessed. This strategy is simpler and more performant compared to the multi-page strategy.	`<a href="reviews.html" data-prefetch>Movie Reviews</a>`
data-rel="back"	This attribute is attached to links or buttons. When set to "back" the link will mimic the back button, going back one history entry (window.history.back()) and ignoring the links default href. For C-Grade browsers (no JavaScript support) the data-rel will be ignored and the href attribute will be used as a fallback (see Listing 2-7).	`<a href="home.html" data-role="button" data-rel="back">Disagree</a>`
data-rel= "dialog"	This attribute is attached to links or buttons. You can set this value to dialog to indicate you want the target page to be styled as a modal dialog (see Listing 2-5).	Terms and Conditions Do you agree to these terms? Disagree Agree `<a href="dialog.html" data-role="button" data-rel="dialog" data-transition="slideup">Show Dialog</a>`

Attribute	Description	Example
data-role= "button"	This attribute is attached to links. Setting the value to button styles a link as a button.	 `<a href="#movies" data-role="button">Show Movies</a>`

data-role= "collapsible"	To create a content block that can expand and collapse, wrap it with an element that contains this attribute (see Listing 6-10).	
		`<div data-role="collapsible">`
		`  <h3>Wireless</h3>`
		`</div>`

data-role= "collapsible-set"	Collapsible sets are similar to collapsible blocks except their collapsible sections are visually grouped together and only one section may be expanded at a time which gives the collapsible set an appearance of an accordion (see Listing 6-11).	
		`<div data-role="collapsible-set">`
		`  <div data-role="collapsible">`
		`    <h3>Wireless</h3>`
		`  </div>`
		`</div>`

data-role= "content"	This attribute is attached to the div element that will contain the content body. This element is optional (see Listing 2-1).	`<div data-role="content">`

data-role= "controlgroup"	If you want to group your buttons together, you can wrap your buttons within a control group (see Listing 4-8).	
		`<div data-role="controlgroup" data-type="horizontal">`
		`<a href="#" data-role="button" class="ui-control-active">In Theatres</a>`
		`<a href="#" data-role="button" class="ui-control-inactive">Coming Soon</a>`
		`</div>`

Attribute	Description	Example
data-role="dialog"	This attribute is attached to the page container. You may apply this attribute as an alternate to data-role="page". Setting the value to dialog will show the page styled as a modal dialog.	<div **data-role="dialog"**>
data-role= "fieldcontain"	This attribute is attached to div elements that wrap form fields. When set to fieldcontain this attribute adds form row separators around the wrapped fields.	<div **data-role="fieldcontain"**> <label><input> </div>
data-role= "footer"	This attribute creates the footer container. The footer is optional (see Listing 2-1).	<div **data-role="footer"**>
data-role= "header"	This attribute creates the header container. The header is optional (see Listing 2-1).	<div **data-role="header"**>
data-role= "list-divider"	This attribute is added to lists to create header segments (see Listing 5-3).	<li **data-role="list-divider"**>Mon</li>
data-role= "listview"	This attribute is used to create list views (see Listing 5-1).	<ul **data-role="listview"**><li><a href="#">Action</a></li></ul>

Attribute	Description	Example
`data-role= "navbar"`	This attribute creates a navigation or tab bar. A navigation bar can be attached to headers or footers (see Listing 3-8).	<div **data-role="navbar">** <ul><li>...</div>
`data-role= "none"`	Adding this attribute to any form or button element tells the framework not to apply any enhanced styles or scripting.	<input type="text" name="name" id="name" value="" **data-role="none"** />
`data-role= "page"`	This attribute defines the page container (see Listing 2-1).	<div data-role="page"> <!-- header --> <!-- content --> <!-- footer --> </div>
`data-role= "slider"`	This attribute is used to create a switch control (see Listing 4-25).	<select name="slider" **data-role="slider">** <option value="off">Off</option> <option value="on">On</option> </select>
`data-shadow`	When false, the framework will remove the drop shadow from the button. Drop shadows will show by default. The "disagree" button shown on the right has its shadow removed.	<a href="" data-role="button" **data-shadow="false"**>Disagree</a>
`data-split-icon`	Sets the icon for the secondary button when building a split button list (see Listing 5-6).	<ul data-role="listview" **data-split-icon="star"**>

Attribute	Description	Example
`data-split-theme`	Sets the theme for the secondary button when building a split button list (see Listing 5-6).	 <ul data-role="listview" data-split-icon="delete" **data-split-theme="v"**>
`data-theme`	This attribute can be added to all containers and page components to create themable designs. For the complete list of available themes refer to Chapter 7.	<div data-role="header" **data-theme="b"**>
`data-title`	This attribute is attached to the page container and sets the title for a page. (see "Setting the Page Title of an Internal Page", Chapter 2).	<div data-role="page" **data-title="Welcome"**>
`data-transition`	This attribute can be attached to links, buttons, and forms. The value of this attribute sets the CSS-based transition effect to use when transitioning between pages. Refer to Table 2-1 for the complete list of available transitions.	<a href="flip.html" **data-transition="flip"**>Flip</a>
`data-type="horizontal"`	Buttons are positioned vertically by default, we can style them horizontally with the addition of the data-type="horizontal" attribute (see Listing 4-8).	 <div data-role="controlgroup" **data-type="horizontal"**> <a href="#" data-role="button" class="ui-control-active">In Theatres</a> <a href="#" data-role="button" class="ui-control-inactive">Coming Soon</a> </div>

Attribute	Description	Example
data-url	This attribute is attached to the page container. The value of this attribute is the pages unique locator and will be shown in the browsers URL bar. By default, jQuery Mobile will assign the unique locator based on the page's URL. However, if you want to change the URL you can set your desired URL as the value of the data-url attribute. For instance, you may want to override the URL after a redirect. Or if you wanted to hide the filename and only show the URL path you could update the data-url attribute to exclude the filename.	<div data-role="page" **data-url**="/override/url/path/or/filename.html">

Summary

In this chapter, we saw how to configure jQuery Mobile and reviewed many of the common API features that are available for building more dynamic pages. Whether you need to globally change the default transition or want to show the back button on all pages, jQuery Mobile allows us to reconfigure many common settings. We also reviewed many popular methods, events, and properties that jQuery Mobile has exposed publicly. These API features are useful when you need to programmatically update your Mobile Web application. Lastly, we saw a complete listing of all jQuery Mobile data attributes. Hopefully, this can listing can provide a quick reference for when you are working on your fast-paced jQuery Mobile projects. Each attribute included a brief description, example, and figure.

In our next chapter, we are going to take an in-depth look at working effectively with services. We will see how to integrate jQuery Mobile pages with client-side and server-side integration solutions.

Chapter 9

Service Integration Strategies

When building Web applications there are two primary access strategies for loading data. There is the traditional, server-side access strategy and the Web 2.0, client-side access strategy. In this chapter, we are going to show examples of integrating jQuery Mobile with both access strategies and we will discuss the advantages of each. jQuery Mobile integrates very well with both strategies and as a result, you can choose the appropriate access strategy that best suits your application needs.

To start, we are going to show two client-side integration examples. With the popularity of social media, our first example will demonstrate how to integrate with Twitter's RESTful API. RESTful APIs are lightweight web services that are often preferred over traditional web services because of their simple setup and flexible response types (JSON, XML). After our Twitter example, we will create our own RESTful API that allows users to register for free movie prizes. This registration example will help demonstrate our ability to POST to a RESTful API from a jQuery Mobile application. In addition, this example will familiarize you with setting up RESTful API's on the server-side.

Then, we will transition into server-side integration strategies and implement one use case that GET's data and another that POST's data. For comparison purposes, we will re-implement our client-side registration example as a server-side solution. You may be surprise to see how much cleaner our page markup becomes when fetching data with a server-side model-view-controller (MVC) access strategy.

Lastly, with the popularity of Geolocation and map views, we will see how to integrate jQuery Mobile with the HTML5 Geolocation API and Google Maps.

Client-side Integration with RESTful Services

Most social media sites have a public API for accessing their data. Twitter,[1] LinkedIn,[2] and Facebook[3] integrations are very common on the Web and RESTful integrations are common access strategies for each of them. In this section, we are going to integrate jQuery Mobile with two different RESTful APIs so we can see how well this strategy runs within jQuery Mobile.

Client-side Twitter Integration with Ajax

In our first client-side example we are going to integrate jQuery Mobile with Twitter's RESTful API. Twitter is a very popular social media site which allows users to send out, or "tweet" brief messages about topics, events, or random opinions. In our movie app, it may be valuable to allow users to search Twitter in real-time for feedback on a movie they might be interested in. For instance, in addition to viewing others' reviews of a particular movie, we may want to provide a convenient link to Twitter that displays the latest tweets about it. On our user review page, we have added a Twitter button to the header that users may click to view the latest tweets about the movie (see Figure 9–1 and its related code in Listing 9–1).

---

[1] See http://dev.twitter.com/console.

[2] See http://developer.linkedin.com/community/apis.

[3] See http://developers.facebook.com/.

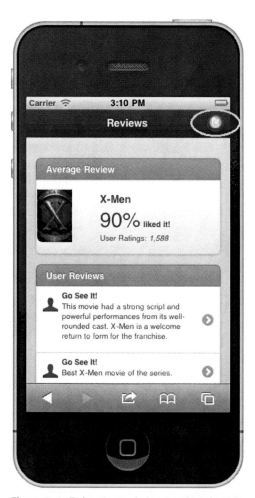

Figure 9–1. *Twitter button in header of movie review page*

Listing 9–1. *Twitter button in header of movie review page (ch9/reviews.html)*

```
<div data-role="page" id="reviewsPage">
    <div data-role="header">
        <h1>Reviews</h1>
        <a href="twitter.html" id="twitterBtn" class="ui-btn-right"
                data-icon="custom" data-iconpos="notext"></a>
    </div>
    ...
</div>
```

When users click the Twitter button, we will search Twitter for the most recent tweets for the current movie and load the results onto our Twitter results page (see Figure 9–2 and its related code in Listing 9–2).

Figure 9–2. *Twitter results page*

Listing 9–2. *Twitter results page (ch9/twitter.html)*

```
<div data-role="page" id="twitterPage">
    ...
    <div data-role="content">
      <ul id="tweet-list" data-role="listview" data-inset="true">
        <li data-role="list-divider">Tweets</p></li>
      </ul>
    </div>
</div>
```

The tweet-list id shown in Listing 9–2 is our placeholder where we will append our Twitter search results. Twitter's search API returns many data elements, however, we are only interested in the tweet text, the user that posted the tweet, and the user's profile image.

TIP: To view all the data elements that are available from Twitter's search API, launch this string in your browser "http://search.twitter.com/search.json?q=xmen". This is the basic search API where the value of the "q" parameter is our searchable keyword(s). In this case, we are searching Twitter for any tweets with the keyword "xmen" in them.

TIP: Most browsers show JSON responses in an unformatted style that can be very unfriendly:

```
{"completed_in":0.063,"max_id":114707386274545
page=2&max_id=114707386274545665&q=xmen","page
2011 14:29:06
+0000","from_user":"am_southpaw","from_user_id
so_language_code":"en","metadata":
{"result_type":"recent"},"profile_image_url":"
```

As an alternative, Firefox has a JSON Viewer plug-in[4] that formats JSON responses in a more structured format:

```
{
    completed_in: 0.086,
    max_id: 1147069989375950850,
    max_id_str: "1147069989375950848",
    next_page: "?page=2&max_id=1147069893759
    page: 1,
    query: "xmen",
    refresh_url: "?since_id=1147069893759508
  - results: [
```

Also, if you need to validate JSON, JSONLint[5] can be a helpful tool.

jQuery Mobile has Ajax support built-in and this makes RESTful integrations simpler with no dependencies on third party JavaScript frameworks. This support comes from the jQuery API[6] that jQuery Mobile extends. In jQuery, the $.ajax[7] API is the preferred

---

[4]   See https://addons.mozilla.org/en-US/firefox/addon/jsonview/.

[5]   See http://jsonlint.com/.

[6]   See http://jquery.com/.

[7]   See http://api.jquery.com/jQuery.ajax/.

solution for RESTful integrations because of its simplicity and its flexible configuration options (timeout, caching, etc.).

To integrate jQuery Mobile with Twitter's RESTful API, the following steps are necessary (see Listing 9–3):

1. When the Twitter button is clicked, we will initially display the jQuery Mobile activity indicator so the user is visually aware that an activity is being processed in the background:

   ```
   $( "#twitterBtn" ).bind( "click", function(e) {
       $.mobile.showPageLoadingMsg();
   ```

2. Next, we will load our Twitter results page into the DOM of the current page. If the page already exists in the DOM we will reload and update the cached page:

   ```
   $.mobile.loadPage("twitter.html", { reloadPage: true });
   ```

3. Before our Twitter page is enhanced, we will send an AJAX request to the Twitter API to gather our search results:

   ```
   $("#twitterPage").live("pagebeforecreate", function(){
       $.ajax({...
   ```

4. Our url option is configured to Twitter's RESTful API and our search query is configured to find all tweets containing the keyword "xmen":

   ```
   url: "http://search.twitter.com/search.json?q=xmen"
   ```

5. Since the Twitter API exists on another domain, we are required to set our dataType option to jsonp. Normally, cross-domain communication is not allowed on the Web but JSONP[8] helps facilitate a trusted means of integration across domains:

   ```
   dataType: "jsonp"
   ```

6. Lastly, we implement our success callback to iterate the search results, create a list item for each row, and append the new markup to our list container:

```
success: function( response ) {...
```

Listing 9–3. *Client-side Twitter integration (ch9/twitter.js)*

```
$( "#reviewsPage" ).live( "pageinit", function(){
    $( "#twitterBtn" ).bind( "click", function(e) {
        $.mobile.showPageLoadingMsg();

        // Reload Twitter results page even if it's already in the DOM
        $.mobile.loadPage("twitter.html", { reloadPage: true });

        // Prevent default click behavior
        return false;
```

---

[8] See http://en.wikipedia.org/wiki/JSONP.

```
    });
});

$( #twitterPage" ).live("pagebeforecreate", function(){
    $.ajax({
        url: "http://search.twitter.com/search.json?q=xmen",
        dataType: "jsonp",
        success: function( response ) {
            // Generate a list item for each tweet
            var markup = "";
            $.each(response.results, function(index, result) {
                var $template = $('<div><li><img class="ui-li-icon profile">
<p class="from"></p><p class="tweet"></p></li></div>');
                $template.find(".from").append(result.from_user);
                $template.find(".tweet").append(result.text);
                $template.find(".profile")
                            .attr("src", result.profile_image_url);
                markup += $template.html();
            });

            // Append the Tweet items into our Tweet list and refresh the
            // entire list.
            $( "#tweet-list" ).append(markup).listview( "refresh", true );

            // Transition to the Twitter results page.
            $.mobile.changePage( $("#twitterPage") );
        },
    });
});
```

In this example, we have chosen to load the Twitter results on demand when the user clicks on the Twitter button. Alternatively, you may pre-fetch the Twitter data so users can see the Twitter results instantly when the button is clicked. To set up this strategy, add the data-prefetch attribute on the Twitter button:

```
<a href="twitter.html" id="twitterBtn" class="ui-btn-right" data-icon="custom" data-
iconpos="notext" data-prefetch></a>
```

Now the page change can be handled by the button's default click behavior allowing us to remove our custom click handler for this button and the $.mobile.changePage() call after the Twitter results are appended to the list.

Also, in a production use case, you will want to configure the timeout and error callback on the $.ajax method to handle any unresponsive or unavailable API's. For instance, if the Twitter API is unresponsive it may be helpful to notify the user:

```
timeout: 6000, // Timeout after 6 seconds
error: function(jqXHR, textStatus, errorThrown) {
    $.mobile.hidePageLoadingMsg();

    // Show error message
    $( "<div class='ui-loader ui-overlay-shadow ui-body-e
      ui-corner-all'><h1>"+ $.mobile.pageLoadErrorMessage +"</h1></div>" )
        .css({ "display": "block", "opacity": 0.96, "top": 100 })
        .appendTo( $.mobile.pageContainer )
        .delay( 800 )
        .fadeOut( 1000, function() {
            $( this ).remove();
```

```
        });
}
```

Client-side Form POST with Ajax

The previous example was a use case that sent a GET request to Twitter's API. GET requests are very common when reading from an API and the `$.ajax` method will default to this type when none is specified. In our next example, we will create our own RESTful API that allows our users to send POST requests. Let's create an API so users can register for prizes. Our user interface will consist of a simple form that only requires an email address (see Figure 9–3 and its related code in Listing 9–4).

Figure 9–3. *Registration form for client-side POST with Ajax*

Listing 9–4. *Registration form for client-side POST with Ajax (ch9/register-client.html)*

```
<div data-role="page" id="registrationPage" data-theme="d">
    <div data-role="header">
      <h1>Registration</h1>
    </div>
```

```
    <div data-role="content">
      <form id="register" method="post">
        <label for="email">Email:</label>
        <input type="email" name="email" id="email" placeholder="Email"
required />
        <input type="submit" id="submit" value="Register" />
      </form>
    </div>
</div>
```

> **TIP:** The input field for the email address includes three newer HTML5 attributes. The
> type="email" field provides two benefits. First, when the field receives focus it will prompt
> the QWERTY keyboard with several useful email keys (see Figure 9–3). Secondly, it will also
> verify that the field contains a valid email address when the form is submitted. For instance, in
> the newer desktop browsers, if a user enters an invalid email they will be prompted with the
> following message:
>
>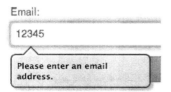
>
> Additionally, the required attribute will assert the email field is not empty when the user
> submits. If it is empty, the user will get this warning:
>
>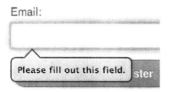
>
> Lastly, the placeholder attribute will add hint text to the input field. While these features are
> helpful, not all of them are supported in today's browsers. Peter-Paul Koch has a useful site[9] that
> shows all available input attributes with their associated browser support.

To handle the form submission on the client-side we will attach an event listener on the
submit button. To submit a POST request to our RESTful API, the following steps are
necessary (see Listing 9–5):

---

[9] See http://www.quirksmode.org/html5/inputs_mobile.html and
http://www.quirksmode.org/html5/inputs.html

1. First, we need to intercept and override the default submit behavior. Now we are ready to submit the form through our RESTful API:

   ```
   $("form").submit(function () {
   ```

2. Secondly, we need to display the jQuery Mobile activity indicator so the user is visually aware that an activity is being processed in the background:

   ```
   $.mobile.showPageLoadingMsg()
   ```

3. Next, we set up our `$.ajax` request with all required options:

   - Our `url` option is configured to our new RESTful resource that was set up locally to handle the client-side registration:

     url: `"http://localhost:8080/jqm-webapp/rest/register"`

     We will look at the RESTful implementation in a moment.

   - Next, we set the `type` option to POST. POST is the recommend type when creating new entities and it is slightly more secure than GET because it does not expose the data attributes as query string parameters on the URL:

     type: `"POST"`

   - Again we set the `dataType` option to `jsonp` because our RESTful API is also running on a separate domain from our client:

     ```
     dataType: "jsonp"
     ```

   - The `jsonp` option defines the callback function that will handle the response:

     ```
     jsonp: "jsoncallback"
     ```

     Any RESTful resource that handles `jsonp` requests must produce a JavaScript response. The response is actually JSON data wrapped within a JavaScript function. For instance, the response of our RESTful API will include the email address that was successfully registered, wrapped in a callback function. The name of this callback function needs to be set as the value of our `jsonp` option:

     jsoncallback(`{"email":"BradPitt@gmail.com"}`)

   - The data option contains the data we want to send to our RESTful resource. In this case, we will send all form values and URL-encode them with jQuery's `serialize` method:

     data: `$("form#register").serialize(),`

▨ The last option is our `success` handler. This will get processed after we receive a successful response from the RESTful API. In our case, we forward the user to a thank you page and also pass the successfully registered email address as a data parameter for confirmation:

```
success: function( response ) {$.mobile.changePage( "register-
thanks.html", {
        data: {"email": response.email}} );
    }
```

Listing 9–5. *Client-side POST with Ajax (ch9/register.js)*

```
$("#registrationPage").live("pageinit", function(){
  $("form").submit(function () {
    $.mobile.showPageLoadingMsg();

    $.ajax({
        url: "http://localhost:8080/jqm-webapp/rest/register",
        type: "POST",
        dataType: "jsonp",
        jsonp: "jsoncallback",
        data: $("form#register").serialize(),
        success: function( response ) {
           $.mobile.changePage( "register-thanks.html",
                { data: {"email": response.email}} );
        }

        return false; // Prevent a form submit
    });
});
```

After a successful registration, users will be forwarded to a thank you page where we show them what they have won along with the email address where the prize was sent (see Figure 9–4 and its related code in Listing 9–6).

Figure 9–4. *Thank you page after client-side POST with Ajax*

Listing 9–6. *Thank you page after client-side POST with Ajax (ch9/register-thanks.html)*

```
<div data-role="page" id="thanksPage" data-theme="d">
  <div data-role="header">
    <a href="/ch9/register-client.html" data-icon="home"  data-iconpos="notext" data-
direction="reverse"></a>
        <h1>Thanks</h1>
  </div>

  <div data-role="content" class="thanks">
    <p>Thanks for registering.  One FREE movie pass was just sent to:    <span
class="email"></span></p>
    <img src="images/free-movie.jpg">
  </div>

</div>
```

> **TIP:** When designing a navigation strategy for your site it is important to always provide the user some navigational option to avoid dead ends. In jQuery Mobile, a simple solution is to always show the home icon in the header bar and have it redirect back to the home page with a reverse transition:
>
> ```
> <a href="home.html" data-icon="home" data-iconpos="notext" data-
> direction="reverse"></a>
> ```

When we executed our changePage call we also passed the email address as a data attribute to the thank you page. That data attribute gets appended to the page's data-url attribute:

```
data-url="/ch9/register-thanks.html?email=BradPitt%40gmail.com"
```

Before the thank you page is enhanced, we fetch that email address and bind it into the email placeholder on our thank you page (see Listing 9–7).

Listing 9–7. *Append email onto thank you page (ch9/register.js)*

```
$("#thanksPage").live("pagebeforecreate", function(){
    var email = getParameterByName("email", $(this).jqmData("url"));
    $(".email").append(email);
});

function getParameterByName(param, url) {
    var match = RegExp('[?&]' + param + '=([^&]*)').exec(url);
    return match && decodeURIComponent(match[1].replace(/\+/g, ' '));
}
```

The RESTful implementation on the server-side to handle the registration was implemented with Jersey[10] and deployed on Tomcat[11] (see Listing 9–8).

Listing 9–8. *RESTful resource to handle registration (com.bmb.jqm.resource.RegisterResourse.java)*

```
@Path("/register")
public class RegisterResource {

    @Produces("application/x-javascript")
    public Response register(@QueryParam("jsoncallback")
        @DefaultValue("jsoncallback") String callback,
        @QueryParam("email") String email) {
        Registration registration = new Registration();
        registration.setEmail(email);
        // Save registration...

        // Return registration in response as jsonp
        return Response.status(Response.Status.OK).entity(new
        JSONWithPadding(registration, callback)).build();
    }
}
```

---

[10] See http://jersey.java.net/.

[11] See http://tomcat.apache.org/.

Let's review Jersey's annotations as we step through the resource:

- The @Path annotation defines the path the resource is responsible for handling. In this case, the RegisterResource object will handle all requests sent to "*/rest/register". Jersey is configured in web.xml and there are two configuration items that require setup (see Listing 9–9). First, we need to define the package(s) where all Resources are deployed, and secondly we need to define what URL patterns should be dispatched through the Jersey container. We have defined that all RESTful resources are declared in package "com.bmb.jqm.resource" and we will route all URL paths with "/rest/*" through the Jersey container.

- The @Produces annotation defines the MIME type of our response. We have chosen to expose our RESTful API publicly across domains which requires the resource to return a JavaScript response. This allows clients to access the API with jsonp requests.

- The register method accepts two input parameters. The first is the callback function name. The client may send the name of the callback function but it is not required. The server will default the callback name to "jsoncallback" if none is supplied. The last parameter is the email address of the user that is registering.

- The server can now process the registration and generate a response. In this example, we will return a response that contains the Registration object converted to JSON and wrapped within the callback function:

```
jsoncallback({"email":"BradPitt@gmail.com"})
```

Listing 9–9. *Jersey configuration (web.xml)*

```
<servlet>
  <servlet-name>Jersey REST Service</servlet-name>
  <servlet-class>
    com.sun.jersey.spi.container.servlet.ServletContainer
  </servlet-class>
  <init-param>
      <param-name>com.sun.jersey.config.property.packages</param-name>
      <param-value>com.bmb.jqm.resource</param-value>
  </init-param>
  <load-on-startup>1</load-on-startup>
</servlet>
<servlet-mapping>
      <servlet-name>Jersey REST Service</servlet-name>
      <url-pattern>/rest/*</url-pattern>
</servlet-mapping>
```

As we have seen, jQuery Mobile integrates very well with RESTful API's. Whether we need to read or submit data, the built-in jQuery library provides all the convenient functions for managing the RESTful lifecycle on the client-side.

Server-side Integration with MVC

In this section, we are going to focus our attention on server-side access strategies. On the Web, a very common strategy is integrating with a model-view-controller (MVC) framework. We will see two MVC examples that will be very similar in style to our client-side examples. In our first example, we will convert our client-side registration use case into a server-side implementation. This example, will provide an apples to apples comparison of how identical use cases can be implemented in jQuery Mobile with client-side versus server-side access strategies. In our final example, we will see how to implement a use case that GET's data from the server.

Server-side Form POST with MVC

For comparison purposes, it will be valuable to see a server-side implementation of our registration use case. Again, we will have a registration form that allows users to opt-in to receive discounted or free movie tickets (see Figure 9–5).

Figure 9–5. *Registration form for server-side POST with MVC*

The page markup for our registration page is very similar to the one shown in our client-side example except we are not going to override the form submission process. In our server-side registration example, when the user clicks the Register button, we will let the form submit its request to our action (see Listing 9–10).

Listing 9–10. *Registration form for server-side integration (/webapp/ch9/register-server.html)*

```
<div data-role="page" id="registrationPage" data-theme="d">
    <div data-role="header">
        <h1>Register</h1>
    </div>

    <div data-role="content">
     <form id="register" action="/jqm-webapp/mvc/register" method="post">
        <label for="email">Email:</label>
        <input type="email" name="email" id="email" placeholder="Email"
        required />

        <input type="submit" value="Register" data-theme="b"/>
    </form>
    </div>
</div>
```

Does anything standout when comparing the registration page from our client-side example (register-client.html) versus this one (register-server.html)? The most notable difference is this page requires no custom JavaScript. As a result, our page markup is much cleaner.

When the form is submitted, a POST request will be sent to the path defined in our action (/jqm-webapp/mvc/register). This request will be handled on the server-side by a Spring MVC[12] controller that is deployed on Tomcat. In our web.xml file, we configured our servlet-mapping so all "/mvc/*" URL's get routed through Spring MVC's dispatcher servlet (see Listing 9–11).

Listing 9–11. *Spring MVC servlet-mapping configuration (/WEB-INF/web.xml)*

```
<servlet>
    <servlet-name>jqm-webapp</servlet-name>
    <servlet-class>
        org.springframework.web.servlet.DispatcherServlet
    </servlet-class>
    <load-on-startup>1</load-on-startup>
</servlet>
<servlet-mapping>
    <servlet-name>jqm-webapp</servlet-name>
    <url-pattern>/mvc/*</url-pattern>
</servlet-mapping>
```

From a user interface perspective, the workflow is identical to our client-side registration example. The form is submitted, processed, and then our thank you page is shown. The

---

[12] See http://static.springsource.org/spring/docs/current/spring-framework-reference/html/mvc.html.

controller code that processes and redirects the request to the thank you page is shown in Listing 9–12.

Listing 9–12. *Spring MVC registration controller (com.bmb.jqm.controller.RegisterController.java)*

```java
@Controller()
public class RegisterController {

    @RequestMapping(method = RequestMethod.POST)
    public String enroll(@RequestParam("email") String email, HttpSession
        session) {
        // Save registration...

        session.setAttribute("email", email);
        return "redirect:/mvc/register/thanks";
    }

    @RequestMapping(method = RequestMethod.GET)
    public String thanks() {
        return "register-thanks";
    }
}
```

Let's review the Spring MVC annotations as we step through the controller:

- The @Controller annotation defines the class as a controller that can handle requests. Spring MVC was configured with path to class name mapping. For instance, all "/register/" requests will be dispatched to the RegisterController. This configuration was setup in Spring MVC's dispatcher servlet (see Listing 9–13).

- The @RequestMapping annotation defines the methods that will handle the POST and GET requests. When the form is submitted, a POST request will be sent to the enroll method. The thanks method will handle all GET requests. For instance, we redirect to the thank you page after the form is processed and the thanks method will get triggered when the thank you page is refreshed.

- The @RequestParam annotation will bind the email address that was submitted on the form to our email input parameter. When the enroll method is called, we save the registration, put the email address in session, and redirect to the thank you page (/jsp/register-thanks.jsp).

Listing 9–13. *Spring MVC path to controller mapping configuration (/WEB-INF/jqm-webapp-servlet.xml)*

```xml
<!-- Enable controller mapping by convention. For example: /foo/* will map to
FooController() -->
<bean
class="org.springframework.web.servlet.mvc.support.ControllerClassNameHandlerMapping" />
```

The appearance of the thank you page will look identical to our client-side example (see Figure 9–6).

Figure 9–6. *Thank you page after server-side POST with MVC*

The only difference is in how the pages are generated. This page is generated on the server-side as a JSP and the email address is bound with JSTL expression syntax (see Listing 9–14). With no JavaScript necessary to dynamically generate the page this markup is cleaner compared to the dynamically generated thank you page we saw in the client-side example.

Listing 9–14. *Thank you page after server-side registration (/jsp/register-thanks.jsp)*

```
<div data-role="page" id="thanksPage" data-theme="d">
    <div data-role="header">
        <h1>Thank You</h1>
    </div>

    <div data-role="content" class="thanks">
        <p>Thanks for registering.  One FREE movie pass was just sent to:
          <span class="email">${email}</span></p>
          <img src="images/free-movie.jpg">
    </div>
</div>
```

One important consideration to be aware of after submitting forms is that jQuery Mobile manages the URL that appears in the browser's location bar. For instance, after the server redirects to "`/mvc/register/thanks`" the browser URL still shows the path of our action ("`/jqm-webapp/mvc/register`"). If you have not implemented a GET request handler for this path, a "refresh" on the thank you page will result in a 404, not found error. You have two options for handling this:

- The simplest solution is to implement a GET request handler on your controller for the action path. Our `RegisterController#thanks` method handles GET requests and will simply refresh the thank you page and redisplay the email address stored in session (see Listing 9–12). Also, when submitting forms on the Web it is recommended to POST and then redirect to avoid any double-submit issues.

- Alternatively, you may manually set the `data-url` attribute on the page container. The value of the `data-url` attribute will be shown in the browser's location bar. This also gives developers more flexibility when constructing semantic paths:

  `data-url="/manually/set/url/path/"`

  This strategy can also be used to hide file names. For instance, if you forward to "`/my/movies/index.html`", you may update the `data-url` attribute of the page to "`/my/movies/`", which will hide the index.html part from displaying.

Server-side Data Access with MVC

In our prior example we saw how to POST form data to the server. In this example, we will use a GET request to fetch data from the server. This example will retrieve a listing of movies from the server and display the results within a jQuery Mobile JSP page (see Figure 9–7).

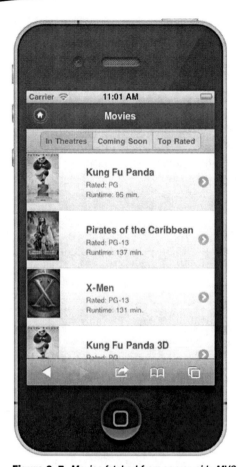

Figure 9–7. *Movies fetched from server-side MVC access*

On the server-side, we have a Spring MVC controller setup to handle GET requests on the following href:

```
<a href="/jqm-webapp/mvc/movies" data-role="button">Movies</a>
```

When the button is clicked, a GET request will be triggered and sent to our MoviesController. The MoviesController will retrieve our movie data and forward the response to the movies JSP page (see Listing 9–15).

Listing 9–15. *MVC Controller to GET movie data (com.bmb.jqm.MoviesController.java)*

```
@Controller()
public class MoviesController {

    @RequestMapping(method = RequestMethod.GET)
    public String getMovies(ModelMap model) {
        model.addAttribute("movies", getMovieData());
        return "movies";
    }
}
```

The response will be forwarded to the movies page where the JSP will iterate the list of movies displaying a separate list item for each result (see Listing 9–16).

Listing 9–16. *JSP to display movie data (/jsp/movies.jsp)*

```
<div data-role="content">
    <ul data-role="listview">
        <c:forEach var="movie" items="${movies}">
        <li>
            <a href="#">
                <img src="../images/${movie.image}" />
                <h3>${movie.title}</h3>
                <p>Rated: ${movie.rating}</p>
                <p>Runtime: ${movie.runtime} min.</p>
            </a>
        </li>
        </c:forEach>
    </ul>
</div>
```

One advantage of this server-side solution is the simplicity of the page markup. There is no dynamic page generation with JavaScript, string concatenation, or dynamic field binding with jQuery selectors that we saw earlier in our client-side examples.

Server-side versus Client-side

Deciding which service access strategy to implement depends on several factors. If you are already building Web applications today you may already have an established pattern for data access on the Web. If so, you may want to continue with this strategy for consistency purposes. Fortunately, when implemented correctly, jQuery Mobile will integrate very well with either strategy and ultimately you can collectively choose which pattern is the best fit for your particular application needs. The following are supporting considerations for each type of strategy:

Client-side integration:

- Faster response times. Client-side integrations produce faster response times because they have fewer point-to-point server dependencies. For instance, we could have aggregated our Twitter data on the server-side but our response times would have decreased due to the additional server communication.

> **CAUTION:** While client-side integrations offer quicker response times, use caution when integrating with third-party API's because it can be advantageous to encapsulate them on the server-side to better isolate your own pages from third-party modifications. For instance, Facebook's RESTful API has changed frequently in the past and now it is actually deprecated.

▨ Faster to implement. Our Twitter example on the client-side was a very quick implementation because only our client-side markup required modifications. Implementing this task on the server side would require client-side and server-side components to be modified.

Server-side integration:

▨ More reliable. A server-side solution is more reliable than a client-side solution because you don't have to be concerned with client-side JavaScript incompatibilities.

▨ More secure. When implementing client-side solutions you must be cautious of the API's and type of data that is exposed. If you are integrating with API's that exposed Personally Identifiable Information (PII), Personal Health Information (PHI), or Payment Card Industry (PCI) information a client-side solution would not be advisable.

▨ Cleaner page markup. We saw examples of this when comparing our pages that were implemented with server-side versus client-side access strategies. The pages used in our server-side access examples had no dependencies on custom JavaScript.

▨ Simpler unit testing of components. I am a believer in the idea that server-side unit testing is still simpler than client-side unit testing. However, after committing several QUnit tests on the jQuery Mobile project, I am beginning to believe that client-side unit testing can be very successful and reliable!

Google Maps Integration

In a recent Mobile Web survey, nearly 75% of Web developers use Geolocation, making it the most popular HTML5 API.[13] When building applications that are location aware, it is common to have a map view displaying points of interest or directions. On the Web, Geolocation[14] in conjunction with Google Maps[15] provide a very useful API for building map functionality. In this section, we are going to see how well jQuery Mobile integrates with Geolocation and Google Maps. To start we will create an example that plots your current location on a map (see Figure 9–8).

---

[13] See http://www.webdirections.org/sotmw2011/.

[14] See http://dev.w3.org/geo/api/spec-source.html.

[15] See http://code.google.com/apis/maps/documentation/javascript/basics.html.

Figure 9–8. *Google Maps integration with jQuery Mobile*

The markup within our page is minimal because we only need to create the content container for our map (see Listing 9–17).

Listing 9–17. *jQuery Mobile page markup for Google Maps integration (ch9/maps.html)*

```
<!DOCTYPE html>
<html>
    <head>
    <meta charset="utf-8">
    <title>Google Maps</title>
    <meta name="viewport" content="width=device-width, initial-scale=1">
    <link rel="stylesheet" type="text/css" href="jquery.mobile.css" />
    <style>
      #map-page, #map-canvas { width: 100%; height: 100%; padding: 0; }
    </style>
    <script type="text/javascript" src="jquery.js"></script>
    <script type="text/javascript" src="maps.js"></script>
    <script type="text/javascript" src="jquery.mobile.js"></script>
    <script src="http://maps.google.com/maps/api/jssensor=false"></script>
</head>
<body>
```

```
<div data-role="page" id="map-page">
    <div data-role="header">
        <h1>Maps</h1>
    </div>

    <div data-role="content" id="map-canvas">
        <!-- map loads here... -->
    </div>
</div>

</body>
</html>
```

When integrating with Google Maps, several additions are necessary:

1. First, we need to style our page and map container so it appears fullscreen:

    ```
    #map-page, #map-canvas { width: 100%; height: 100%; padding: 0; }
    ```

2. Next, we import custom JavaScript to help determine the user's geolocation and draw our map view. We will look at the details of this file shortly:

    ```
    <script type="text/javascript" src="maps.js"></script>
    ```

3. Then, we import the Google Maps API:

    ```
    <script src="http://maps.google.com/maps/api/jssensor=false">
    ```

4. Lastly, we identify our map container. Our map will be drawn within this element:

    ```
    <div data-role="content" id="map-canvas">
    ```

The custom JavaScript to help determine the user's geolocation and draw our map view is shown in Listing 9–18.

Listing 9–18. *JavaScript for Google Maps integration (ch9/maps.js)*

```
$( "#map-page" ).live( "pageinit", function() {

    // Default to Hollywood, CA when no geolocation support
    var defaultLatLng = new google.maps.LatLng(34.0983425, -118.3267434);

    if ( navigator.geolocation ) {
        function success(pos) {
            // Location found, show coordinates on map
            drawMap(new google.maps.LatLng(
                pos.coords.latitude, pos.coords.longitude));
        }

        function fail() {
            drawMap(defaultLatLng);  // Show default map
        }

        // Find users current position
        navigator.geolocation.getCurrentPosition(success, fail,
            {enableHighAccuracy:true, timeout: 6000, maximumAge: 500000});
```

```
    } else {
        drawMap(defaultLatLng); // No geolocation support
    }

    function drawMap(latlng) {
        var myOptions = {
            zoom: 10,
            center: latlng,
            mapTypeId: google.maps.MapTypeId.ROADMAP
        };

        var map = new google.maps.Map(
            document.getElementById("map-canvas"), myOptions);

        // Add an overlay to the map of current lat/lng
        var marker = new google.maps.Marker({
            position: latlng,
            map: map,
            title: "Greetings!"
        });
    }
});
```

When the map page is in a "ready" state we will perform the following steps to draw our map view:

1. First, we determine if the browser supports Geolocation:

   ```
   if ( navigator.geolocation ) {
   ```

2. If the browser supports Geolocation we will attempt to retrieve the users current position:

   ```
   navigator.geolocation.getCurrentPosition(success, fail,
   ```

   ```
   {enableHighAccuracy:true, timeout: 6000, maximumAge: 500000});
   ```

   The getCurrentPosition API can take up to three parameters. The first parameter is the success callback. This is the only required parameter. The next parameter is the error callback and the final parameter is our configurable options. We have configured our Geolocation lookup to use high accuracy. This will attempt to use GPS for positioning if supported. We also configured the timeout to 6 seconds. If we fail to find the users position after 6 seconds our error callback will be invoked. Lastly, we have configured that the lookup may use a cached position if it is less than 5 minutes old.

3. When a successful position is found the success callback will be invoked. In this case, we will draw our map using the position coordinates:

   ```
   function success(pos) {
       drawMap(new google.maps.LatLng(
           pos.coords.latitude, pos.coords.longitude))
       }
   ```

4. Lastly, the `drawMap` method will draw a Google Map with an overlay icon appearing in the center of the map at the position that was identified as your location. However, if Geolocation is not supported or if no position was established, Hollywood, CA will be shown as the default location (see Listing 9–18).

Summary

In this chapter we saw how to integrate jQuery Mobile with client-side and server-side data access strategies. jQuery Mobile integrates very well with both strategies allowing you to choose the access method that is most appropriate for your application needs. While the client-side strategy offers better performance and can be faster to implement it may not be as reliable, secure, or as maintainable as a server-side alternative.

Lastly, we saw an example of how to integrate jQuery Mobile with Geolocation and Google Maps. With these two mapping API's, we now have the ability to add map views to our jQuery Mobile applications.

In Chapter 10, we will look at how we can take our existing jQuery Mobile applications and distribute them natively with PhoneGap.

Easy Deployment with PhoneGap

Native apps appear to have two distinct advantages when compared to Mobile Web applications. First, native apps can be distributed in an app store. The most notable app stores include Apple's App Store, Android Market, HP App Catalog, BlackBerry App World, and Windows Marketplace. App stores simplify the user experience when consumers need to search, purchase, install, or rate native applications. Another advantage native apps have is their ability to interact with device APIs. For instance, native applications have the ability to communicate with most device APIs including contacts, calendar, camera, and the network API to name a few.

In this chapter, we will discuss how we can break these Mobile Web barriers. In particular, we will introduce PhoneGap and show how PhoneGap can help bridge these gaps for our jQuery Mobile apps. As an example, we will take an existing jQuery Mobile app, wrap it with PhoneGap and deploy our app to the native iOS and Android platforms.

We will also see how we can distribute our jQuery Mobile apps to an app store without PhoneGap. For instance, Open App Market is an app store for HTML5 mobile apps that can be an alternative for those that find the native app store distribution process cumbersome and slow.

Lastly, we will take a peek at the progress the W3C is making on client-side device APIs that browsers will someday support. This will be very important for Mobile Web because it will allow our Web applications to access device APIs (calendar, contacts, camera, etc.) with zero dependencies on external frameworks.

What is PhoneGap?

PhoneGap[1] is an open-source development framework that allows you to build cross-platform native apps with web technologies like jQuery Mobile. For instance, we can take an existing jQuery Mobile web app, wrap it with the PhoneGap framework and distribute it to all native platforms that PhoneGap supports. Currently, PhoneGap supports the native iOS, Android, BlackBerry, webOS, and Symbian platforms. In addition to PhoneGap's native distribution capabilities, it also exposes an API that allows our Mobile Web applications to interact with device specific APIs including the file system, notifications, and camera to name a few. For the complete list, refer to PhoneGap's supported features by platform.[2] The PhoneGap API allows us to extend our jQuery Mobile applications in ways that were previously only possible with native SDK's.

Running jQuery Mobile as an iOS App

In this section, we are going to wrap a jQuery Mobile app with PhoneGap and run it on the native iOS platform. To set up PhoneGap for the iOS platform, we can reference PhoneGap's "Getting Started Guide with iOS."[3] PhoneGap has step-by-step instructions for installing PhoneGap on every platform and their instructions are very detailed with screenshots for assistance. Installation of Xcode, the IDE for iOS development[4] is a prerequisite for developing to the iOS platform. If you choose to bypass the Xcode installation it will still be valuable to follow along to familiarize yourself with the general steps that are necessary to set up PhoneGap on a native platform. While each platform has specific IDE setup instructions, the general process of installing PhoneGap, setting up the project, and deploying are consistent steps for all platforms. After your iOS platform is set up, you should have a new Xcode project that looks similar to Figure 10–1.

---

[1] See http://www.phonegap.com/.

[2] See http://www.phonegap.com/about/features.

[3] See http://www.phonegap.com/start#ios-x4.

[4] See http://developer.apple.com/xcode/.

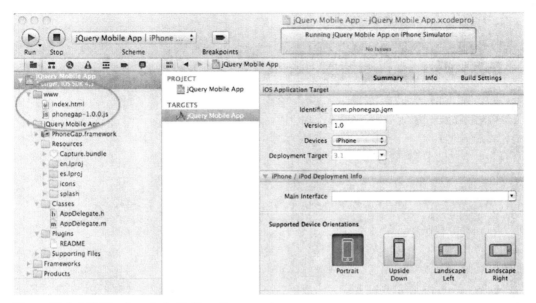

Figure 10–1. *Initial Xcode project with PhoneGap support*

The "www" directory shown in Figure 10–1 is the application root directory. Within this directory are the PhoneGap JavaScript library and a default page (index.html). The index.html page will be shown as the initial landing page when we run the app. In Xcode, to build and run the app, click the "Run" button that appears in Xcode's upper left corner. After clicking "Run", the app will compile, the iOS simulator will launch, and the index page will be shown (see Figure 10–2).

Figure 10–2. *Initial screen when running PhoneGap's default app in Xcode*

With a PhoneGap project setup in Xcode we can now import an existing jQuery Mobile app into our project. The steps for importing a jQuery Mobile app into our Xcode project and deploying as a native iOS app are listed below:

1. First, we need to import an existing jQuery Mobile project into Xcode's "www" root directory. For this exercise, you may import your own jQuery Mobile app or import the jQuery Mobile project that is included in the chapter 10 source code folder. For example, if we import the jQuery Mobile files from the Chapter 10 source code directory and move them into our "www" directory, our Xcode project structure should appear as the figure below:

2. After importing the jQuery Mobile project into our PhoneGap project we need to import PhoneGap's JavaScript library as a top-level resource:

```
<head>
  <meta charset="utf-8">
  <title>jMovies</title>
  <meta name="viewport" content="width=device-width, initial-scale=1">
  <link rel="stylesheet" type="text/css" href="jquery.mobile-min.css" />
  <link rel="stylesheet" type="text/css" href="custom.css" />
  <script type="text/javascript" src="phonegap-1.0.0.js"></script>
  <script type="text/javascript" src="jquery-min.js"></script>
  <script type="text/javascript" src="custom.js"></script>
  <script type="text/javascript" src="jquery.mobile-min.js"></script>
</head>
```

The PhoneGap library is an API that provides access to many device specific features (camera, media, storage, etc.). PhoneGap has documentation and examples of all their supported APIs on their website[5]. Importing the PhoneGap library is only necessary when your application needs to interact with PhoneGap's native capabilities.

3. The last step is to run and test our application. In Xcode, click the "Run" button. This will compile the app and launch it within the iOS simulator. If you imported the jQuery Mobile project from the chapter 10 source code the initial screen to appear will be the springboard (see Figure 10–3).

---

[5] See http://docs.phonegap.com/.

Figure 10–3. *jQuery Mobile running as a native iOS app*

To help validate that the PhoneGap library was properly installed I added a listener for PhoneGap's device-ready event. When this event fires, PhoneGap is in a ready state and we may begin communicating with the PhoneGap API (see Listing 10–1).

Listing 10–1. *PhoneGap is ready (ch10/custom.js)*

```
$(document).bind("deviceready", function(){
    navigator.notification.alert("PhoneGap is initialized...");
});
```

As shown in Figure 10–3, when PhoneGap is in a ready state we display an alert view indicating that PhoneGap has been initialized. The alert notification in Listing 10–1 is an example of how we can programmatically interact with PhoneGap's API to access native functionality.

> **NOTE:** PhoneGap has simplified the process of converting our jQuery Mobile web app to a native platform running on iOS. From a technical perspective, our jQuery Mobile web app is now running within an iOS Web View.

Do you see any differences when comparing our jQuery Mobile app running within a Safari browser (see Figure 10–4) versus the native app running in iOS (see Figure 10–5)?

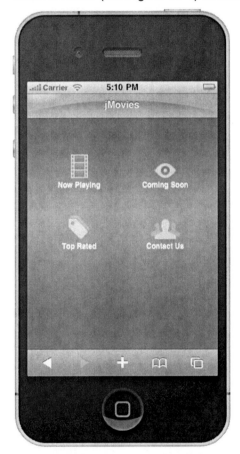

Figure 10–4. *jQuery Mobile running within a browser*

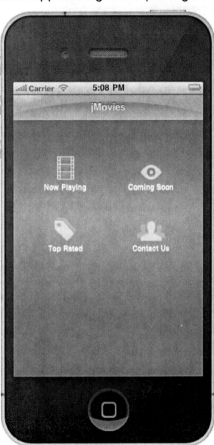

Figure 10–5. *jQuery Mobile running as a native iOS app*

The most obvious difference is that the browser chrome is available within the Safari browser, but not in the native app. If you recall our "Back Button" Section from Chapter 3, back buttons are initially disabled in jQuery Mobile because the browser chrome already provides build-in navigation buttons. However, for users running native iOS apps, the back button within the header is a primary means of navigation. Fortunately, we can enable back buttons in jQuery Mobile with a simple configuration update (see Listing 10–2).

Listing 10–2. *Globally enable back buttons (ch10/custom.js)*

```
$(document).bind("mobileinit", function(){
    $.mobile.page.prototype.options.addBackBtn = true;
});
```

The back button in jQuery Mobile is very intelligent. It will only appear when there is a page in history to go back to. After enabling the back button globally our native iOS users will feel more comfortable when navigating the app (see Figure 10–6).

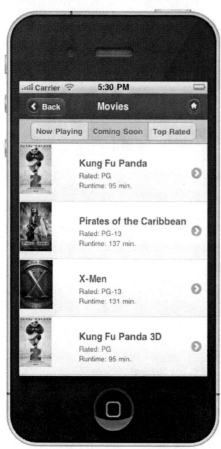

Figure 10–6. *Globally enable back buttons for navigation support*

With back buttons enabled globally we may also want to disable them on specific pages where they are not necessary. In particular, we will want to prevent the back button from appearing on the home screen. To prevent the back button from appearing on a given page, we can add the `data-add-back-btn="false"` attribute to the page container:

```
<div data-role="page" id="home" data-add-back-btn="false">
```

As a result, when we navigate back to our home screen the back button will not be shown.

Now that we are capable of deploying our jQuery Mobile app to the native iOS platform we will also want to customize the default app icon (see Figure 10–7) and splash screen (see Figure 10–8).

Figure 10–7. *PhoneGap's default app icon in iOS*   **Figure 10–8.** *PhoneGap's default splash screen in iOS*

App icons are stored in the project's /Resources/icons directory and splash screen images are stored in the /Resources/splash directory (see Figure 10–9). Images are available for different iOS screen densities and sizes.

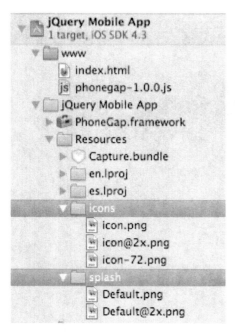

Figure 10–9. *Xcode's images for splash screen and app icons*

Also, when you need to change the bundle display name or identifier, those can be set in Xcode in the project's info tab (see Figure 10–10). The bundle display name sets the label for the app icon and the bundle identifier is used by iOS to uniquely identify your application.

PROJECT	Summary	Info	Build Settings
![jQuery Mobile App] jQuery Mobile App	▼ Custom iOS Target Properties		

Key	Type	Value
Localization native development region	String	English
Bundle display name ‡ ⊙ ⊙	String	jMovies
Executable file	String	${EXECUTABLE_NAME}
Icon file	String	icon.png
▶ Icon files	Array	(3 items)
Bundle identifier	String	com.phonegap.jqm

TARGETS

jQuery Mobile App

Figure 10–10. *Setting bundle display name and identifier*

> **TIP:** When developing with PhoneGap, it is recommended to set the `$.mobile.allowCrossDomainPages` configuration option to `true`:
>
> ```
> $(document).bind("mobileinit", function(){
>     $.mobile.allowCrossDomainPages = true;
> });
> ```
>
> Phone Gap's web view allows applications to make cross-domain calls. This is usually allowed so the application can fetch data from their home server. By default, jQuery Mobile will treat cross-domain requests as external links. As a result, the cross-domain page will not be loaded into the DOM of the current page and no transitions will be applied. Therefore, if you want to allow jQuery Mobile to manage the page loading logic of cross-domain requests in PhoneGap, set this option to `true`.

That completes the entire process from installing PhoneGap to running our jQuery Mobile app on the native iOS platform. After your app is production ready, the final step is distributing your iOS app to Apple's App Store. Although the process can be lengthy, the complete instructions for distributing your app to Apple's App Store can be found in Apple's iOS developer library[6].

Running jQuery Mobile as an Android App

In this section, we are going to wrap a jQuery Mobile app with PhoneGap and run it on the native Android platform. To set up PhoneGap on the Android platform, we will reference PhoneGap's "Getting Started Guide with Android"[7]. Installation of Eclipse, the IDE for Android development, is a prerequisite. After your Android platform is set up, you should have a new Eclipse project that looks similar to Figure 10–11.

---

[6] See http://developer.apple.com/library/ios/#documentation/Xcode/Conceptual/ios_development_workflow/145-Distributing_Applications/distributing_applications.html.

[7] See http://www.phonegap.com/start#android.

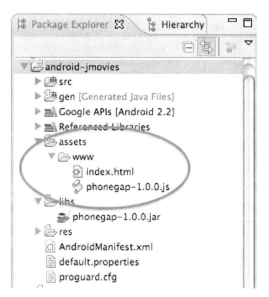

Figure 10–11. *Initial Eclipse project with PhoneGap support*

The "www" directory shown in Figure 10–11 is the application root directory. Within this directory are the PhoneGap JavaScript library and a default page (index.html). The index.html page will be shown as the initial landing page when we run the app. In Eclipse, to build and run the app, click the Run menu, select Run As, and choose Android Application. After we compile and run the app, the Android simulator will launch and the index page will be shown (see Figure 10–12).

Figure 10–12. *Initial screen when running PhoneGap's default app in Eclipse*

TIP: If you find it takes too long for the Android simulator to launch you may prefer to deploy to an actual device for testing. For this setup, make sure USB debugging is enabled on your device (**Settings ➤ Applications ➤ Development**) and plug it into your system. Now when you run your application, it will launch much quicker.

With a PhoneGap project setup in Eclipse we can now import an existing jQuery Mobile app into our project. The steps for importing a jQuery Mobile app into our Eclipse project and deploying as a native Android app are listed below:

1. First, we need to import an existing jQuery Mobile project into Eclipse's "www" root directory. For this exercise, you may import your own jQuery Mobile app or import the jQuery Mobile project that is included in the chapter 10 source code folder. For example, if we import the jQuery Mobile files from the Chapter 10 source code directory and move them into our "www" directory, our Eclipse project structure should appear as shown in Figure 10–13:

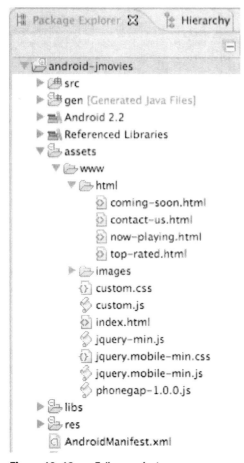

Figure 10–13. *an Eclipse project*

2. After importing the jQuery Mobile project into our Eclipse project we need to import PhoneGap's JavaScript library as a top-level resource:

```
<head>
  <meta charset="utf-8">
  <title>jMovies</title>
  <meta name="viewport" content="width=device-width, initial-scale=1">
  <link rel="stylesheet" type="text/css" href="jquery.mobile-min.css" />
  <link rel="stylesheet" type="text/css" href="custom.css" />
  <script type="text/javascript" src="phonegap-1.0.0.js"></script>
  <script type="text/javascript" src="jquery-min.js"></script>
```

```
<script type="text/javascript" src="custom.js"></script>
<script type="text/javascript" src="jquery.mobile-min.js"></script>
</head>
```

3. The last step is to run and test our application. In Eclipse, click the Run menu, select Run As, and choose Android Application. This will compile the app and launch it within the Android simulator. If you imported the jQuery Mobile project from the chapter 10 source code the initial screen to appear will be the springboard (see Figure 10–14).

Figure 10–14. *Initial screen when running PhoneGap from Eclipse*

If you recall from our iOS example we globally enabled the back button to appear on all screens because back buttons are often fixed within the header of an iOS app. Since Android has a hardware based back button, it is not necessary to enable our back buttons (see Figure 10–15).

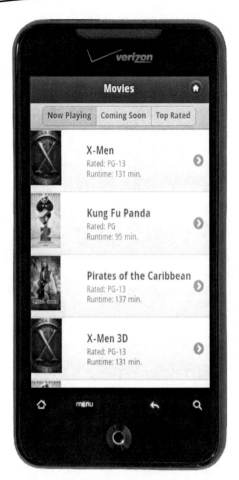

Figure 10–15. *No back button necessary on Android*

> **TIP:** Extract platform specific dependencies in separate files. This separation will help simplify the management of different configurations across platforms. For instance, create a separate configuration file for each supported platform: custom-ios.js, custom-android.js. Now each platform can load their unique dependencies.

Now that we are capable of deploying our jQuery Mobile app to the native Android platform we will also want to customize the default app icon (see Figure 10–16).

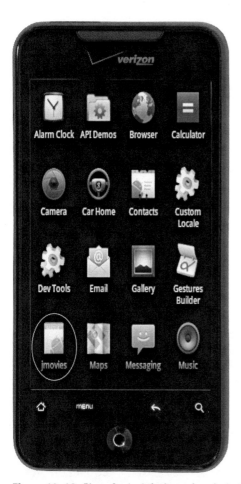

Figure 10–16. *PhoneGap's default app icon in Android*

The Android app icons are stored in the project's /res/drawable-* directories with images available for high, medium, and low densities (see Figure 10–17).

Figure 10–17. *Android's app icon images*

That completes the entire process from installing PhoneGap to running our jQuery Mobile app on the native Android platform. When your app is ready for production, the final step is distributing your Android app to Android Market. The complete instructions for distributing your app to Android Market can be found in Android's Development Guide.[8]

Open App Market

Open App Market[9] is an app store for HTML5 mobile apps that allows us to search, purchase, install, or rate HTML5 mobile apps just like the native app stores (see Figure 10–18).

---

[8] See http://developer.android.com/guide/publishing/publishing.html.

[9] See http://openappmkt.com/.

Figure 10–18. *Open App Market*

To get started with Open App Market we must first install it onto our device. Currently, Open App Market is available to iOS and Android users. To install Open App Market, scan the QR code in Figure 10–19 with your iOS or Android device.

Figure 10–19: *QR code to install Open App Market. Alternatively, you can go to openappmkt.com and click the installation link at the top of their site.*

After installing the app, you can search for Mobile Web apps by category or popularity. When you find a free or paid app you are interested in, download it and the app will be saved on your home screen just like an app from a native app store. For instance, Figure 10–20 shows the Open App Market app alongside the Twitter and YouTube apps that were downloaded from Open App Market.

Figure 10–20. *Open App Market downloads apps in the same manner as native stores*

Client-side Device APIs

If you need to build a mobile app that must integrate with device specific features like camera, contacts, or the network, what mobile technology are you going to choose? Today, our choices are limited. We must build with either a native platform or use a hybrid technology like PhoneGap. It would be ideal if all web browsers had support for these device specific features too. While no browsers have support for these features today, the W3C is currently implementing working drafts for most of the major client-side device APIs[10]. The most notable client-side device APIs include access to camera,

---

[10] See http://www.w3.org/2009/dap/.

network, calendar, contacts, messaging, and battery information. While it may be too soon to predict when browsers will have support for these features, at least progress is well underway.

Summary

In this chapter we saw how to take an existing jQuery Mobile application and integrate it with the PhoneGap framework. PhoneGap adds two unique capabilities to jQuery Mobile web applications. First, we can take our existing jQuery Mobile web apps, wrap them within the PhoneGap framework and distribute them to the native iOS, Android, BlackBerry, WebOS, and Symbian platforms. Secondly, PhoneGap also exposes an API that allows our Mobile Web applications to interact with device specific APIs, including the file system, notifications, camera, and many more. The PhoneGap API allows us to extend our jQuery Mobile applications in ways that were previously only possible with native SDK's.

We were also introduced Open App Market, an app store for HTML5 mobile apps that allows us to search, purchase, install, or rate HTML5 mobile apps just like the native app stores. The Open App Market can be an alternative for those that find the native app store distribution process cumbersome and slow.

Lastly, we introduced the client-side device APIs that the W3C is currently authoring. These APIs will be very important for Mobile Web developers because it will allow our Web applications to access device APIs (calendar, contacts, camera, etc.) with zero dependencies on external frameworks.

Index

CPSIA information can be obtained at www.ICGtesting.com
Printed in the USA
LVOW051757280312

275165LV00005B/5/P